INDUSTRIAL APPLICATIC
BOUNDARY ELEMENT ME. ....

Developments in Boundary Element Methods—5

PREVIOUS TITLES IN THIS SERIES

Developments in Boundary Element Methods—1
Edited by P.K. Banerjee and R. Butterfield (1979)
Advanced application to a wide range of problems in engineering

Developments in Boundary Element Methods—2
Edited by P.K. Banerjee and R.P. Shaw (1982)
Applications to time-dependent and time harmonic problems

Developments in Boundary Element Methods—3
Edited by P.K. Banerjee and S. Mukherjee (1984)
Advances in boundary element analysis of non-linear problems of solid and fluid mechanics

Developments in Boundary Element Methods—4
Edited by P.K. Banerjee and J.O. Watson (1986)
A range of advanced engineering problems

INDUSTRIAL APPLICATIONS OF BOUNDARY ELEMENT METHODS

Developments in Boundary Element Methods—5

Edited by

P.K. BANERJEE

Department of Civil Engineering,
State University of New York at Buffalo, USA

and

R.B. WILSON

Structures Technology Division,
Pratt & Whitney, Hartford, Connecticut, USA

Routledge
Taylor & Francis Group

LONDON AND NEW YORK

First published 1989 by Elsevier Science Publishers Ltd

2 Park Square, Milton Park, Abingdon, Oxfordshire OX14 4RN
52 Vanderbilt Avenue, New York, NY 10017

Routledge is an imprint of the Taylor & Francis Group, an informa business

First issued in paperback 2019

British Library Cataloguing in Publication Data

Industrial applications of boundary element methods.
  1. Engineering. Mathematics. Boundary element methods
  I. Banerjee, P.K. (Prasanta Kumar) II. Wilson, R.B.
  III. Series
  620′.001′515353

  ISBN 1-85166-356-8

Library of Congress Cataloging-in-Publication Data
(Revised for vol. 5)

Developments in boundary element methods.

  Includes bibliographies and index.
  Contents: v. 5. Industrial applications of
boundary elements methods/edited by P.K. Banerjee
and R.B. Wilson.
  1. Boundary element methods. I. Banerjee, P.K.
(Prasanta Kumar), 1941- . II. Wilson, R.B. (Raymond
Bruce), 1942-
TA347.B69D48 1985   620′.001′51535   85-7072
ISBN 0-85334-376-4 (v. 4)

Photoset by Interprint Ltd (Malta)

ISBN 978-1-85166-356-9 (hbk)
ISBN 978-0-367-86358-6 (pbk)

PREFACE

The development of the boundary element method (BEM) as a general
problem solving tool, which started in the early 1960s, has now reached
a fairly advanced state. It is now possible to attempt the solutions of realistic
and complex problems of the engineering industry. During the last 12 years,
many large general purpose codes have been written and distributed. One
of the pioneering efforts in this regard was that of Lachat and Watson who
basically laid the foundations for this type of development. This volume of
the series has been specifically designed to show engineering applications of
some of these codes together with sufficient exposition of the theoretical
background, so that a prospective code developer or a user can gain the
necessary insight into the current status of BEM.

Chapter 1 describes the development and use of the code BEST3D in the
elastic and inelastic analysis of gas turbine engine structures. The very first
large scale implementations of elastoplastic analyses of two-dimensional,
three-dimensional and axisymmetric problems are discussed in Chapter 2,
where analyses have also been developed through the method of particular
integrals in which volume integration was found to be unnecessary.
Chapter 3 discusses the application of BEM to problems of periodic and
transient dynamic analyses. The general implementation of both steady
state and transient coupled problems of thermoelasticity and poroelasticity
is described in Chapter 4, where uncoupled thermoelasticity has been
presented as a special case of the general theory. All of these implementations described in Chapters 1 to 4 are parts of the GPBEST system. Some of
the three-dimensional analyses are also included in the NASA funded

programme BEST3D, which is available to the public. Advanced development of BEM in dynamic fracture mechanics analysis is described in Chapter 5. Chapter 6 describes the use of the code CRX3D (derived from BEST3D) for multi-region (substructured) fracture mechanics analysis. Chapter 7 documents the history of the use of BEM in the US automotive industry. It appears that BEM is not a very popular analysis tool for automotive components. The use of GPBEST for the acoustic eigen-frequency analysis is outlined; it has been observed that BEM is an extremely efficient tool for this acoustic analysis. Chapter 8 describes the development and use of the code DBETSY for large three-dimensional stress analyses. The size of the problems analysed makes these probably among the largest applications of BEM to date. Finally Chapter 9 describes the implementation of stress analysis and heat transfer in the code EZBEA for the analysis of tractor components.

We hope that the developments described here will convince the reader that BEM can now be used in routine solutions of complex problems of the engineering industry.

P.K. BANERJEE
R.B. WILSON

CONTENTS

Preface v

List of Contributors ix

1. Advanced Applications of BEM to Gas Turbine Engine
   Structures 1
   R.B. WILSON, N. MILLER, and P.K. BANERJEE

2. Advanced Applications of BEM to Inelastic Analysis of Solids   39
   P.K. BANERJEE and D.P. HENRY, JR

3. Advanced Development of BEM for Elastic and Inelastic
   Dynamic Analysis of Solids 77
   P.K. BANERJEE, S. AHMAD and H.C. WANG

4. Boundary Element Methods for Poroelastic and Thermoelastic
   Analyses 119
   G.F. DARGUSH and P.K. BANERJEE

5. Advanced Substructured Analysis of Elastodynamic Wave
   Propagation Problems 157
   N. NISHIMURA, S. KOBAYASHI and M. KITAHARA

6. BEM Analysis of Problems of Fracture Mechanics 187
   S.T. RAVEENDRA and T.A. CRUSE

7. Boundary Element Applications in the Automotive Industry... 205
   W.E. FALBY and P. SURULINARAYANASAMI

8. Advanced Stress Analysis by a Commercial BEM Code....... 231
   H.J. BUTENSCHÖN, W. MÖHRMANN and W. BAUER

9. Thermoelastic Analysis for Design of Machine Components... 263
   GERALD L. GRAF and YOSEPH GEBRE-GIORGIS

Index.. 295

LIST OF CONTRIBUTORS

S. AHMAD
Department of Civil Engineering, State University of New York at Buffalo, Buffalo, New York 14260, USA

P.K. BANERJEE
Department of Civil Engineering, State University of New York at Buffalo, Buffalo, New York 14260, USA

W. BAUER
Daimler-Benz AG, Postfach 60 02 02, 7000 Stuttgart 60, FRG

H.J. BUTENSCHÖN
Daimler-Benz AG, Postfach 60 02 02, 7000 Stuttgart 60, FRG

T.A. CRUSE
Southwest Research Institute, Department of Engineering Mechanics, 6220 Culebra Road, San Antonio, Texas 78284, USA

G.F. DARGUSH
Department of Civil Engineering, State University of New York at Buffalo, Buffalo, New York 14260, USA

W.E. FALBY
Ford Motor Company, Car Products Division, 20000 Rotunda Drive, Dearborn, Michigan 48121, USA

YOSEPH GEBRE-GIORGIS
Caterpillar Tractor Company, Research Department, Technical Center, Peoria, Illinois 61629, USA

GERALD L. GRAF
Caterpillar Tractor Company, Research Department, Technical Center, Peoria, Illinois 61629, USA

D.P. HENRY, JR
Department of Civil Engineering, State University of New York at Buffalo, Buffalo, New York 14260, USA

M. KITAHARA
Department of Ocean Engineering, Tokai University, Shimizu, Japan

S. KOBAYASHI
Department of Civil Engineering, Kyoto University, Kyoto 606, Japan

N. MILLER
Structures Technology Division, Pratt & Whitney, Hartford, Connecticut 06108, USA

W. MÖHRMANN
Daimler-Benz AG, Postfach 60 02 02, 7000 Stuttgart 60, FRG

N. NISHIMURA
Department of Civil Engineering, Kyoto University, Kyoto 606, Japan

S.T. RAVEENDRA
Southwest Research Institute, Department of Engineering Mechanics, 6220 Culebra Road, San Antonio, Texas 78284, USA

P. SURULINARAYANASAMI
Ford Motor Company, Car Products Division, 20000 Rotunda Drive, Dearborn, Michigan 48121, USA

H.C. WANG
Department of Civil Engineering, State University of New York at Buffalo, Buffalo, New York 14260, USA

R.B. WILSON
Structures Technology Division, Pratt & Whitney, Hartford, Connecticut 06108, USA

Chapter 1

ADVANCED APPLICATIONS OF BEM TO GAS TURBINE ENGINE STRUCTURES

R.B. Wilson, N. Miller

Structures Technology Division, Pratt & Whitney, Hartford, Connecticut, USA

and

P.K. Banerjee

Department of Civil Engineering, State University of New York at Buffalo, USA

SUMMARY

The boundary element method (BEM) has been known for some time to be extremely useful for the solution of elastic stress analysis problems involving high stress/strain gradients. In particular, the method has been extensively used for the study of both two- and three-dimensional fracture mechanics problems. Recent analytical and numerical developments coupled with the general availability of greatly increased computing capacity have made both elastic and inelastic three-dimensional stress analysis feasible for complex geometries such as those found in gas turbine engine components.

This chapter summarises the features of an advanced three-dimensional stress analysis code based on BEM for elastic, inelastic and vibration analyses of multizone or substructured three-dimensional solids. The elastic analyses have been developed for isotropic or cross-anisotropic media with thermal and centrifugal loading. The inelastic analyses include isotropic plasticity with variable hardening and kinematic plasticity with multiple yield surfaces. Free-vibration analysis has only been developed for the isotropic three-dimensional solid.

1

1.1 INTRODUCTION

The mathematical background of the boundary element method has been known for nearly 100 years. Indeed, some of the boundary integral formulations for elastic, elastodynamic, wave and potential flow equations have existed in the literature for at least 50 years. With the emergence of digital computers the method had begun to gain popularity as 'the boundary integral equation method', 'panel method', 'integral equation method', etc., during the 1960s. The name was changed to 'Boundary Element Method' by Banerjee and Butterfield (1975), so as to make it more appealing to the engineering analysis community. Since then a number of textbooks and advanced level monographs have appeared which give a very comprehensive and thorough account of the existing literature on the method. The development of BEM has now reached a stage when attempts can be made to analyse realistic complex engineering problems.

A method intended for the three-dimensional stress analysis of gas turbine engine components must meet rather stringent demands, based on the nature of the problems to be solved. Some of the characteristics of these problems are briefly discussed below.

1.1.1 Geometry

Gas turbine engine structures, particularly the hot section components (turbine blades, turbine vanes and burner liners), are geometrically complex. This results from the need simultaneously to minimise weight, maintain efficient aerodynamics, deliver required cooling air and achieve acceptable part durability. Intersecting cavities or holes of various types are frequently encountered, as are surfaces with reverse curvature. Since it is often the case that the major motivation for BEM analysis is the calculation of accurate stresses and strains for use in durability analysis, the proper representation of these geometrical details, in addition to overall part geometry, is crucial.

1.1.2 Problem Size

Although turbine engine components are small in size, relative to civil engineering structures, the geometric complexity of the structures leads to large problem sizes, even if quite advanced methods are used for the representation of the geometry. A reasonable estimate for a high turbine blade model would be 400 to 800 quadratic surface elements, involving as many as 7200 degrees of freedom. The integral nature of the method leads to a full matrix in the final algebraic representation of the problem. Even with present computing capability it is clear that the efficient solution of such problems, particularly with inelastic response, requires extensive substructuring capability.

Further, it often happens that the geometry or loading for part of a structure will be modified during the design or development process, while the remainder of the part is unaltered. An effective substructuring and restart capability is required to accommodate such changes.

Finally, it is clear that, with current or projected computing capabilities, it is not possible to analyse a complex turbine engine component in full detail in a single analysis. For cost-effective analysis it is necessary to be able to work with a refined, and highly accurate, mesh in the regions of primary interest while using a much coarser model in remote locations. It is, therefore, highly desirable to provide a method of locally improving the stresses without rebuilding the mesh.

1.1.3 Loading Definition

The loading in turbine engine problems is of three main types: surface mechanical loads, centrifugal loads and thermal loads, normally due to non-steady temperature fields. The surface loads are most frequently tractions, although on some occasions displacements (usually rollers) are applied. It is often desirable to apply these conditions in a local coordinate system.

Centrifugal loads are present for all the rotating components (blades, discs, sideplates and spacers) and frequently provide the major portion of the mechanical loads.

Thermal loads are due to the presence of non-constant, non-steady temperature fields. These loads can be significant even for rotating hardware and are the dominant source of load in non-rotating parts such as turbine vanes and burner liners.

1.1.4 Material Property Definition

The materials used in hot section components typically have temperature-dependent elastic material properties. In addition, in turbine blades, anisotropic materials are often used. These parts are cooled after casting to produce material with all grain boundaries in the same direction (directionally solidified material) or without grain boundaries (single crystal material). While these are relatively simple types of anisotropy, they add considerable complexity to the BEM analysis since the kernel function becomes much more complex.

In addition to the elastic property definition it is also necessary to describe the inelastic response of the material. All hot section components experience at least local plasticity. In turbine blades and discs this is normally limited to the immediate vicinity of notched details or to the disc bore. In burner liners and in the aerofoil section of turbine vanes plastic and creep behaviour often influence the entire component.

It is clear from the foregoing discussion that a stress analysis tool for such components must have very advanced capability. Unfortunately all existing boundary element implementations suffer from a lack of either generality or an acceptable level of accuracy, even for a general substructured elastic system, and as such cannot be applied to even the elastic analysis of the present problem. Moreover, to date, the inelastic analysis has only been implemented for simple test problems involving single regions. Free-vibration problems, on the other hand, have never before been attempted using a three-dimensional BEM formulation.

The purpose of this chapter is therefore to demonstrate that for the first time BEM formulations have been developed to analyse extremely complex and large problems such as those existing in the gas turbine engine environment. In addition new advanced applications of BEM to problems of monotonic and cyclic plasticity and to problems of free-vibration are described. Some of the more advanced features such as anisotropy, thermally sensitive directonally solidified non-linear material modelling, etc., are in their concluding stages of development and will be described in the very near future.

1.2 BOUNDARY INTEGRAL FORMULATIONS

1.2.1 Elastic Analysis with Body Forces
The elastic analysis of solids subjected to conservative body forces can be very efficiently developed by using the theory of particular integrals. The method of particular integrals has been tentatively discussed by Banerjee and Butterfield (1981) and Lachat and Watson (1976). Recently the method was applied to the treatment of body forces resulting from self-weight and centrifugal loading by Pape and Banerjee (1987), Henry et al. (1987) and Banerjee et al. (1988a).

The method of particular integrals is well known in the solution of inhomogeneous differential equations. For example, a linear inhomogeneous differential equation of the form

$$L(u_i) + \psi_i = 0 \tag{1}$$

in which $L(u_i)$ is a self-adjoint homogeneous differential operator and ψ_i is the inhomogeneous known quantity, can be solved by this method. Since this is a linear equation, its solution can be represented as the sum of a complementary solution u_i^c satisfying

$$L(u_i^c) = 0 \qquad (2)$$

and a particular integral u_i^p satisfying

$$L(u_i^p) + \psi_i = 0 \qquad (3)$$

The total solution is then simply

$$u_i = u_i^c + u_i^p \qquad (4)$$

The following observations are important as far as the particular integrals are concerned:

(1) They can often be obtained by inspection of the inhomogeneous differential equation or by the method of undetermined coefficients.
(2) There are no unique particular integrals, since any polynomial satisfying eqn (3) is a valid particular integral.

For gravitational and centrifugal loading one can work out particular integrals as illustrated below.

1.2.2 Gravitational Loading
If we consider the case of gravitational acceleration directed along the x_3-axis, the body force components are given as

$$\psi_1 = \psi_2 = 0 \qquad (5)$$

$$\psi_3 = -\rho g \qquad (6)$$

in which ρ is the density and g is the acceleration due to gravity. Using the method of undetermined coefficients, we can find a particular solution of eqn (3) for this set of body forces. The three-dimensional form of these particular integrals is then:

$$u_1^p = \frac{-\lambda \rho g}{2\mu(3\lambda + 2\mu)} x_1 x_3 \qquad (7)$$

$$u_2^p = \frac{-\lambda \rho g}{2\mu(3\lambda + 2\mu)} x_2 x_3 \qquad (8)$$

$$u_3^p = \frac{(\lambda + \mu)\rho g}{2\mu(3\lambda + 2\mu)} x_3^2 + \frac{-\lambda \rho g}{4\mu(3\lambda + 2\mu)} (x_1^2 + x_2^2) \qquad (9)$$

1.2.3 Centrifugal Loading

The body forces corresponding to the rotation of a body about the x_3-axis of a cartesian coordinate system are given by:

$$\psi_1 = \rho\omega^2 x_1 \tag{10}$$

$$\psi_2 = \rho\omega^2 x_2 \tag{11}$$

$$\psi_3 = 0 \tag{12}$$

where ω is the angular velocity, and ρ is again the density of the material. A set of particular integrals for this loading is:

$$u_1^p = \frac{-\rho\omega^2}{8(\lambda + 2\mu)} \left[\frac{5\lambda + 4\mu}{4(\lambda + \mu)}(x_1^2 + x_2^2) + \frac{\mu}{\lambda + \mu} x_3^2 \right] x_1 \tag{13}$$

$$u_2^p = \frac{-\rho\omega^2}{8(\lambda + 2\mu)} \left[\frac{5\lambda + 4\mu}{4(\lambda + \mu)}(x_1^2 + x_2^2) + \frac{\mu}{\lambda + \mu} x_3^2 \right] x_2 \tag{14}$$

$$u_3^p = \frac{\rho\omega^2}{8(\lambda + 2\mu)}(x_1^2 + x_2^2)x_3 \tag{15}$$

Once u_i^p is obtained, the solution can be developed by constructing the conventional BEM formulation of the homogeneous differential equation of elasticity, i.e.

$$C_{ij}u_i^c(\xi) = \int_s [G_{ij}(x, \xi)t_i^c(x) - F_{ij}(x, \xi)u_i^c(x)] \, ds \tag{16}$$

which when discretised takes the form

$$Gt^c - Fu^c = 0 \tag{17}$$

This can then be converted using a particular integral to give

$$Gt - Fu = Gt^p - Fu^p \tag{18}$$

or

$$Ax = b + Gt^p - Fu^p = b + b^p \tag{19}$$

The traction t^p associated with u^p can be obtained from the strain displacements and stress–strain relations. The kernel functions G_{ij} and F_{ij} are given in Banerjee and Butterfield (1981).

1.2.4 Inelastic Analyses

The basic integral representation for the elastoplastic problem (Banerjee *et al.*, 1979; Banerjee and Cathie, 1980; Banerjee and Butterfield, 1981) can be expressed as

$$c_{ij}\dot{u}_i(\xi) = \int_s [G_{ij}(x, \xi)\dot{t}_i(x) - F_{ij}(x, \xi)\dot{u}_i(x)] \, ds$$

$$+ \int_v B_{ikj}(x, \xi)\dot{\sigma}^0_{ik}(x) \, dv \qquad (20)$$

where c_{ij} is the second-rank tensor derived from the treatment of the singular integral involving the F_{ij} kernel as in the elastostatics case, as the field point ξ moves to the boundary. For an interior point $c_{ij} = \delta_{ij}$.

The stress rate at an interior point ξ is obtained from eqn (20) by using the constitutive relationships $(\dot{\sigma}_{ij} = D^e_{ijkl}\dot{\varepsilon}_{kl} - \dot{\sigma}^0_{ij})$ as (Banerjee and Davies, 1984; Banerjee and Raveendra, 1986):

$$\dot{\sigma}_{jk}(\xi) = \int_s [G^\sigma_{ijk}(x, \xi)\dot{t}_i(x) - F^\sigma_{ijk}(x, \xi)\dot{u}_i(x)] \, ds$$

$$+ \int_v B^\sigma_{ipjk}(x, \xi)\dot{\sigma}^0_{ip}(x) \, dv + J^\sigma_{ipjk}\dot{\sigma}^0_{ip}(\xi) \qquad (21)$$

All the kernel functions G^σ, F^σ and B^σ as well as the jump term J^σ are defined in Banerjee and Davies (1984) and Banerjee and Raveendra (1986).

The expressions for strains and stresses cannot be constructed at the boundary points by taking the field point (ξ) in eqn (3) to the surface point, due to the strongly singular nature of the integrals involved. The evaluation of strains and stresses at boundary points can be accomplished by considering the equilibrium of the boundary segment and utilising constitutive and kinematic equations. The stresses and global derivatives of the displacements which lead to strains at point ξ can be obtained by coupling the following set of equations:

$$\dot{\sigma}_{ij}(\xi) - [\Delta\delta_{ij}\dot{u}_{k,k}(\xi) + \mu[u_{i,j}(\xi) + u_{j,i}(\xi)]] = -\dot{\sigma}^0_{ij}(\xi)$$

$$\dot{\sigma}_{ij}(\xi)n_j(\xi) = \dot{t}_i(\xi) \qquad (22)$$

$$\frac{\partial\xi_k}{\partial\eta_l}\frac{\partial u_i}{\partial\xi_k} = \frac{\partial u_i}{\partial\eta_l}$$

where η_i are a set of local axes at the field point ξ, and Δ and μ are Lame's constants.

By making use of the Lagrangian shape functions in the third equation of (22) and rearranging into known and unknown components, the above sets of equations can be inverted and rearranged to form

$$\dot{\sigma}_{jk} = \bar{G}^{\sigma}_{ijk}\dot{t}_i - \bar{F}^{\sigma}_{ijk}\dot{u}_i + \bar{B}^{\sigma}_{ipjk}\dot{\sigma}^{o}_{ip}$$

$$\dot{\varepsilon}_{jk} = \bar{G}^{\varepsilon}_{ijk}\dot{t}_i - \bar{F}^{\varepsilon}_{ijk}\dot{u}_i + \bar{B}^{\varepsilon}_{ipjk}\dot{\sigma}^{o}_{ip}$$

(23)

Equations (20), (21) and (23) can be solved together either iteratively (Banerjee et al., 1979; Banerjee and Butterfield, 1981) or using a direct solution algorithm (Raveendra, 1984; Banerjee and Raveendra, 1987).

1.2.5 Free-vibration Analyses

Free-vibration analysis capability by BEM has always been severely limited because the periodic formulation has the frequency parameter non-linearly embedded in the kernel functions G and F. Nardini and Brebbia (1982) and Ahmad and Banerjee (1986) developed algorithms that make use of the static fundamental solutions such that the final eigenvalue extraction is reduced to a generalised eigenvalue problem. Both these pairs of authors express the inertia term via global shape functions and use either the divergence theorem (Nardini and Brebbia, 1982) or particular integrals (Ahmad and Banerjee, 1986; Wang and Banerjee, 1988; Banerjee et al., 1988b) to convert the problem into a boundary only category. The method developed by Ahmad and Banerjee (1986) has been developed as described below for three-dimensional free-vibration problems.

The governing differential equation for free-vibrations of an elastic, homogeneous and isotropic body can be written as

$$(\lambda + \mu)\frac{\partial^2 u_j}{\partial x_i \partial x_i} + \mu\frac{\partial^2 u_i}{\partial x_j \partial x_j} + \rho\omega^2 u_i = 0$$

(24)

where λ and μ are Lame's constants.

Equation (24) can also be written in operator notation as

$$L(u_i) + \rho\omega^2 u_i = 0$$

(25)

The solution of the above equation can be represented as the sum of a complementary function u_i^c satisfying

$$L(u_i^c) = 0$$

(26)

and a particular integral u_i^p satisfying

$$L(u_i^p) + \rho\omega^2 u_i = 0$$

(27)

However, eqn (27) still contains the unknown displacement field u_i within the domain, which can be eliminated by using an unknown fictitious density function ϕ and a known function C, exactly as in an indirect boundary element analysis. More specifically

$$u_i(\mathbf{x}) = \sum_{m=1}^{\infty} C_{ik}(\mathbf{x}, \xi^m)\phi_k(\xi^m) \tag{28}$$

where ϕ_k is a fictitious density, and C_{ik} is a known function which can be selected as any linear function of spatial coordinates.

A simple function which is selected for C_{ik} in the present analysis has the form (Nardini and Brebbia, 1982; Ahmad and Banerjee, 1986):

$$C_{ik}(\mathbf{x}, \xi^m) = \delta_{ik}(R - r) \tag{29}$$

where:

R = largest distance between two points of the body
r = distance between $\mathbf{x}$ (field point) and ξ^m (source point).
On the basis of the preceding assumption, eqn (27) can be written as

$$L(u_i^p) + \rho\omega^2 \sum_{m=1}^{\infty} C_{ik}(\mathbf{x}, \xi^m)\phi_k(\xi^m) = 0 \tag{30}$$

Now, the particular integral u_i^p can be chosen as any function which satisfies the differential eqn (30). Accordingly it can be represented as:

$$u_i^p(\mathbf{x}) = \sum_{m=1}^{\infty} D_{ik}(\mathbf{x}, \xi^m)\phi_k(\xi^m) \tag{31}$$

The displacement field satisfying eqn (30) is found to be

$$D_{ik}(\mathbf{x}, \xi^m) = \frac{\rho\omega^2}{\mu}\left[(c_1 r - c_2 R)\delta_{ik}r^2 - c_3 y_i y_k r\right] \tag{32}$$

where:

$$y_i = x_i - \xi_i^m$$

$$c_1 = \frac{2(d+3)(1-v)-1}{18(3d-1)(1-v)}$$

$$c_2 = \frac{1-2v}{2[(1+d)-2vd]}$$

$$c_3 = \frac{1}{2(1-v)(d^2+4d+3)}$$

v = Poisson's ratio

d = dimensionality of the problem
(e.g. for three-dimensional problems, $d=3$).

The surface traction t_i^p related to the displacement u_i^p can be determined using the strain–displacement relationship and constitutive equation and is given by:

$$t_i^p(\mathbf{x}) = \sum_{m=1}^{\infty} T_{ik}(\mathbf{x}, \boldsymbol{\xi}^m)\phi_k(\boldsymbol{\xi}^m) \tag{33}$$

where:

$$T_{ik}(\mathbf{x}, \boldsymbol{\xi}^m) = \rho\omega^2 [(c_4 r - c_5 R)y_k n_i + (c_6 r - 2C_2 R)y_i n_k + \{(c_6 r - 2c_2 R)\delta_{ik} - 2c_3 y_i y_k/r\}y_j n_j] \tag{34}$$

where:

$$c_4 = \frac{(d+3)v-1}{3(3d-1)(1-v)}$$

$$c_5 = \frac{2v}{(1+d)-2vd}$$

$$c_6 = \frac{(d+2)-(d+3)v}{3(3d-1)(1-v)}$$

The boundary values of real displacements and tractions u_i and t_i can be related to the complementary and particular integrals via

$$u_i = u_i^c + u_i^p \tag{35}$$

$$t_i = t_i^c + t_i^p \tag{36}$$

By usual discretisation of the boundary S, we can express the complimentary solution in matrix form as

$$[G]\{t^c\} - [F]\{u^c\} = \{0\} \tag{37}$$

Equation (37) can be solved once the displacements u_i^c and tractions t_i^c are expressed in terms of the real displacement u_i and traction t_i, i.e.

$$[G]\{t\} - [F]\{u\} = [G]\{t^p\} - [F]\{u^p\} \tag{38}$$

where vectors $\{t^p\}$ and $\{u^p\}$ can be obtained at boundary nodes from eqns (32) and (33) as

$$\{u^p\} = \rho\omega^2[D]\{\phi\} \tag{39}$$

$$\{t^p\} = \rho\omega^2[T]\{\phi\} \tag{40}$$

Substituting these equations into eqn (38), we obtain

$$[G]\{t\} - [F]\{u\} = \rho\omega^2([G][T] - [F][D])\{\phi\} \tag{41}$$

Recalling that

$$u_i(\mathbf{x}^n) = \sum_{m=1}^{\infty} \delta_{ij}(R - r^{nm})\phi_j(\xi^m) \tag{42}$$

where r^{nm} is the distance between the points $\mathbf{x}^n$ and ξ^m, we can express this relationship between the displacements and the fictitious density at all boundary nodes as:

$$u_i^n = \delta_{ij}p^{nm}\phi_j^m \tag{43}$$

$$\phi_j^m = \delta_{ij}K^{nm}u_i^l \tag{44}$$

where

$$K^{nm} = (P^{nm})^{-1}$$

We can now write eqn (44) as

$$\{\phi\} = [K]\{u\} \tag{45}$$

Substituting $\{\phi\}$ from eqn (45) into eqn (41) we get

$$[G]\{t\} - [F]\{u\} = \rho\omega^2([G][T] - [F][D])[K]\{u\} \tag{46}$$

or

$$[G]\{t\} - [F]\{u\} = \rho\omega^2[M]\{u\} \tag{47}$$

Equation (47) can also be written in terms of known and unknown variables as

$$[A]\{x\} - [B]\{y\} = \rho\omega^2([\bar{M}]\{x\} - [M^*]\{y\}) \tag{48}$$

Since all the known variables are zero (i.e. specified boundary conditions are either $u_i = 0$ or $t_i = 0$), eqn (48) reduces to

$$[A]\{x\} = \rho\omega^2[\bar{M}]\{x\} \tag{49}$$

The modified mass matrix $[\bar{M}]$ contains zero in its sub-columns related to specified displacements (i.e. fixed boundaries).

1.3 NUMERICAL IMPLEMENTATION

The integral equation formulation discussed in the preceding section becomes of practical interest only when numerical techniques are employed for its solution, since analytical solutions exist only for the simplest geometries and loading.

Boundary integral formulations for any problem represent the exact statement of the problem posed. In order to maximise the benefits, it is essential not to introduce any unnecessary approximations purely for ease of programming. In this regard, the authors avoid the use of non-conforming elements. Although the use of such elements reduces skilled programming effort by nearly an order of magnitude, it becomes virtually impossible to analyse any realistic three-dimensional problem by such a system. Indeed, a pilot study to compare the relative efficiency and accuracy of conforming and non-conforming elements (Manolis and Banerjee, 1986) shows that conforming elements are more accurate for the same number of quadratic elements, and that computing costs are much higher for non-conforming elements. The authors were much inspired by the work of Lachat and Watson (1976) who described the first implementation of BEM for the elastic analysis of large three-dimensional structures. The present implementation is essentially a more efficient, accurate and general version of that type of implementation with additional capabilities of inelastic, dynamic and free-vibration analyses of a multi-region substructured three-dimensional system.

In basic terms, the outline of the numerical solution process is as follows:

(1) Approximate the problem geometry and the variation of field quantities (displacement, traction, strain, etc.) using a suitable set of computationally convenient simple functions.
(2) Evaluate the surface and volume integrals numerically.
(3) Assemble a set of algebraic equations by utilising the boundary conditions specified for the problem.
(4) Solve the equation system.

In the case of a transient problem (or a quasi-static analysis with multiple loads) steps 3 and 4 are repeated. In a non-linear analysis, iteration at each load or step will normally be required in order to satisfy the constitutive equation (Banerjee and Raveendra, 1986; Banerjee et al., 1987).

The basic steps in the numerical solution process are very briefly discussed in the sections which follow. The method employed for the approximation of both geometry and field quantities is based on the use of the isoparametric shape functions originally developed for use in finite element analysis. The entire boundary of the part is modelled as the union of a set of six or eight noded elements. Over each element the variation of each of the cartesian coordinates is approximated as

$$x_i = N^k(\eta)x_i^k \tag{50}$$

where the x_i^k are the nodal values of the coordinate and the N^k are interpolation functions which take the value 1·0 at single node and vanish at all others. Continuity along element boundaries is ensured.

In addition to the element types mentioned above, elements which extend to infinity are provided. These elements are designed to allow modelling of structures connected to ground, and automatically incorporate appropriate decay conditions. The characteristics of the various element types are summarised in Table 1.

TABLE 1

Element type	Geometry nodes	Displacement/ traction nodes
Linear quadrilateral	8	4
Linear triangle	6	3
Quadratic quadrilateral	8	8
Quadratic triangle	6	6
Linear infinite	8	2
Quadratic infinite	8	3

All the surface integrals in the numerical implementations have been calculated numerically. Since this is the most time-consuming portion of the analysis it is essential to optimise this effort. Essentially two types of integrals, singular and non-singular, are involved. The integrals are singular if the source point for the equations being constructed lies on the element being integrated. Otherwise, the integrals are non-singular, although numerical evaluation is still difficult if the source point and the element being integrated are close together. A self-adaptive segmentation based on error control has been devised for these integrals.

In both the singular and non-singular cases Gaussian integration is used. The basic technique was developed in Banerjee and Butterfield (1981) and was first applied in the three-dimensional boundary element method by Lachat and Watson (1976). This allows the determination of element subdivisions and orders of Gaussian integration which will retain a consistent level of error throughout the structure. Numerical tests have shown that the use of 3×3, 4×4 and 5×5 Gauss rules provide the best combination of accuracy and efficiency. In the present implementation the Gaussian rules (2×2 for low, 3×3 for medium and 4×4 for high accuracy) are used for non-singular integration, and error is controlled through element subdivision. The origin of the element subdivision is taken to be the closest point to the source point on the element being integrated (Fig. 1).

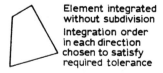

Element integrated
without subdivision

Integration order
in each direction
chosen to satisfy
required tolerance

o

Case 1   Source point remote from element

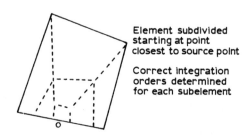

Element subdivided
starting at point
closest to source point

Correct integration
orders determined
for each subelement

Case 2   Source point near element

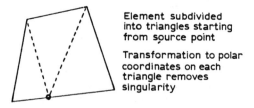

Element subdivided
into triangles starting
from source point

Transformation to polar
coordinates on each
triangle removes
singularity

Case 3   Source point on element

FIG. 1   Self-adaptive integration scheme.

If the source point is very close to the element being integrated, the use of a uniform subdivision of the element can lead to excessive computing time. This frequently happens, due either to mesh transitions or to the analysis of thin-walled structures. In order to improve efficiency while retaining accuracy, a graded element subdivision was employed. Based on one-dimensional tests, it was found that the subelement divisions could be allowed to grow geometrically away from the origin of the element subdivision. Numerical tests on a complex three-dimensional problem have shown that a mesh expansion factor and aspect ratio as high as 4·0 can be employed without significant degradation of accuracy.

In the case of singular integration (source point on the element being integrated) the element is first divided into triangular subelements. The integration over each subelement is carried out in a polar coordinate system with origin at the source point. This coordinate transformation produces non-singular behaviour in all except one of the required integrals. Normal Gauss rules can then be employed. The remaining integral (that of the traction kernel F_{ij} times the isoparametric shape function which is 1·0 at the source point) is still singular, and cannot be numerically evaluated with reasonable efficiency and accuracy. Its calculation is carried out indirectly, using the concept of rigid body motion. It has been found that subdivision in the circumferential direction is required to preserve accuracy in the singular integration. A maximum included angle of 15 degrees is used. Subdivision in the radial direction has not been required.

Volume modelling is also based on the use of the isoparametric shape functions. Both volume cell geometry and the variation of field quantities (initial stress and temperature) within the cell are mapped to a $2 \times 2 \times 2$ cube using the quadratic isoparametric shape functions. Volume modelling is required in the boundary element method only when thermal loads, elastic inhomogeneity and/or non-linear material response are involved, and only the portion of the part in which these effects occur must be modelled. Nodes of the volume cells lying on the surface need not match nodes of the surface mesh in either location or node numbering.

The intermediate result of the surface and volume integrations is a set of coefficients which function as multipliers of field quantities (displacements, tractions, strains, stresses, temperatures). Some of the field quantities are known from the boundary conditions. During the assembly of the equation system the known and unknown variables are separated and expressed in appropriate local coordinate systems. The coefficients multiplying each set of variables are collected in matrix form for later use. Boundary conditions are imposed, including any required modifications to the coefficient matrices. The results of these operations in matrix form for all the systems are

$$Ax = By + R \qquad (51)$$

and

$$\sigma = A^{\sigma}x + B^{\sigma}y + R^{\sigma} \qquad (52)$$

where:

$x =$ vector of unknown displacements and tractions
$y =$ vector of known displacements and tractions
$R, R^{\sigma} =$ vectors of body force and initial stress contributions.

In any substructured problem, the matrix A in eqn (51) contains large blocks of zeros, since separate regions communicate only through common surface elements. In order to save both storage space and computer time the matrix is stored in a block basis, the zero blocks being ignored. The matrices in eqn (52) are block diagonal, since interior results in any region involve only surface and volume integrations relative to that region.

In a non-linear analysis the constitutive equations of the form

$$\dot{\sigma}_{ij}(\xi) = D^{ep}_{ijkl}\dot{\varepsilon}_{kl} \qquad (53)$$

with

$$\dot{\sigma}^{o}_{ij}(\xi) = (D^{e}_{ijkl} - D^{ep}_{ijkl})\dot{\varepsilon}_{kl}$$

defining the initial stress rates for an elasto-plastic material and the form

$$\dot{\sigma}^{o}_{ij}(\xi) = D^{vp}_{ijkl}\sigma_{kl} \qquad (54)$$

defining the initial stress rate for a viscoplastic material are necessary to establish the magnitude of the vector R^{σ} through an iterative process (Banerjee et al., 1979; Banerjee and Butterfield, 1981). This algorithm to establish the initial stress has been accelerated by utilising the past history of the initial stress development (Banerjee and Raveendra, 1986; Banerjee et al., 1987; Henry and Banerjee, 1987).

The routine used for the solution of the generalised algebraic eigenvalue problem is based on the work of Saad (1984). The original paper addressed the extraction of the largest few eigenvalues of a very large, very sparse system arising in the development of multi-grid methods. Two processes, an iteration and a purification step, were used in the original method. To date, only the iterative process has been required in the BEST3D application. In order to employ the algorithm in BEST3D it was necessary to reformulate it for the generalised eigenvalue problem and adapt it to the block storage used in BEST3D.

The generalised eigenvalue problem:

$$(A - \omega^2 B)x = 0 \tag{55}$$

in which the lowest eigenvalues are sought, is reformulated as the ordinary eigenvalue problem:

$$(A^{-1}B - \lambda I)x = 0 \tag{56}$$

with

$$\lambda = 1/\omega^2$$

in which the largest few eigenvalues are sought. Starting from an arbitrary initial vector, x^0, successive vectors orthogonal to one another with respect to the matrix $A^{-1}B$ are formed. At each step of the iteration an upper Hessenberg matrix is formed, whose eigenvalues converge to the eigenvalues of the original system. Convergence is excellent, with the first mode usually obtained in five to eight iterations. Computational experience has shown that one additional mode per iteration is normally obtained thereafter. It should be noted that the inversion implied in eqn (56) is not necessary and it is only the decomposed form that is required.

1.4 EXAMPLES

The analysis described above has been applied to a very large number of three-dimensional components related to gas turbine engine structures (Banerjee et al., 1985; Wilson et al., 1985; Wilson and Banerjee, 1986). A few representative examples are described below. Since the analysis described above has been incorporated in a general-purpose computer code known as BEST (Boundary Element Solution Technique), the results of the present analysis have been designated as the 'BEST3D' analysis.

1.4.1 C-notch Low Cycle Fatigue Specimen
The C-notch low cycle fatigue specimen, as shown in Fig. 2, was designed to place a large volume of material in a plane strain, high stress condition. The specimen has, since its design several years ago, been subjected to a variety of elastic and inelastic analyses as well as to strain gauge testing for specimen calibration.

In the present verification programme the specimen was analysed using BEST3D. The portion of the specimen which was analysed is indicated in Fig. 2. The mesh used is shown, in both full and hidden line views, in Fig. 3.

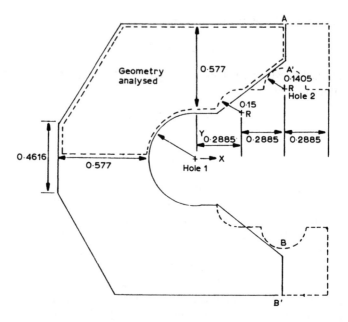

FIG. 2   Cross-section of C-notch low cycle fatigue specimen. (All measurements in
inches. 1 inch = 2·54 cm.)

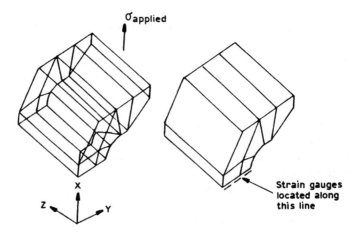

FIG. 3   BEST model for analysis of C-notch low cycle fatigue specimen.

The analysis was carried out using both linear and quadratic variations of displacement and traction.

Key stress results are summarised in Table 2. The plane strain results were obtained from a variety of two-dimensional codes. The baseline results were obtained from a very detailed three-dimensional analysis. The NASTRAN results cited were obtained using a mesh of 20 node isoparametric elements. The surface mesh refinement in the NASTRAN analysis was approximately equivalent to that in the BEST3D analysis.

TABLE 2

	Plane strain	Baseline	NASTRAN	BEST3D quadratic	BEST3D linear
σ_{xx} (midplane)	250	252	224	248	191
σ_{xx} (free surface)	—	150	136	158	138
σ_{zz} (midplane)	76	75	67	70	—

It is clear from the results in Table 2 that BEST3D (using quadratic variation) is equivalent in accuracy to the previous baseline solution and to the plane strain results, and is superior to three-dimensional finite element analysis for an equivalent mesh.

It is also clear that the use of linear variation for the full model does not provide sufficiently accurate results. The linear results could be improved by mesh refinement, but the use of quadratic variation over the same mesh is more efficient in both input preparation and analysis cost.

1.4.2 Benchmark Notch Specimen

The benchmark notch specimen is a double edge notch specimen developed by General Electric/Louisiana State University (GE/LSU) under NASA-Lewis Contract NAS3-22522. A significant volume of well-documented data was provided in the final report (Domas et al., 1982). These data can be used to verify both the elastic and inelastic capabilities of the present analysis.

The specimen geometry is defined in Fig. 4. Stress analysis was carried out for the gauge section only (because of the interest in obtaining the stress concentration factor), a procedure already known to be satisfactory. The material involved in the specimen was forged Inconel 718, a nickel-base superalloy commonly used in gas turbine engine components.

For the elastic analysis ($E = 23\cdot8 \times 10^6$ ksi, $v = 0\cdot3$) linear, quadratic and

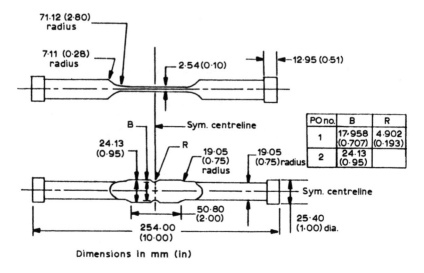

FIG. 4   Double edge notch specimen.

a mixture of linear and quadratic surface elements were used. The major characteristics of all these analyses are described in Table 3.

It can be seen that most of the better analyses in Table 3 give values of the maximum peripheral strain (at an average axial stress of 100 ksi) at the free surface of the notch between 1700 and 1800 microstrain, which agrees with average experimental data. It is also clear that the analysis involving linear variation of displacements and tractions over the surface elements cannot provide acceptable stress concentrations without unacceptable mesh

TABLE 3

Model	Elements	Displacement and traction variation	Equations	Subregions	IBM3081 CPU time (seconds)	Max. strain
1	50	linear	156	1	64	1 688
2	50	quadratic	456	1	234	1 780
3	22	linear	78	2	20	1 594
4	22	quadratic	210	2	60	1 742
5	22	mixed	117	2	31	1 729
6	22	linear	36	1	10	1 186
7	10	quadratic	96	1	28	1 605

refinement. Models 2, 4 and 5 all yield results of accuracy entirely consistent with the experimental data. The variation in peak strain among these three is within $\pm 1.5\%$ of the mean value.

The most important observation from these analyses is that the most cost-effective analysis (5) uses substructuring (two regions) with mixed linear and quadratic surface elements. The surface element discretisation for this case is shown in Fig. 5.

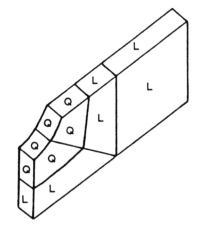

L Linear variation
Q Quadratic variation

FIG. 5   Optimised model for analysis of benchmark notch specimen.

In this model eight of the 22 elements (those in and near the notch) were quadratic, while the remaining 14 were linear. The visible elements in the hidden line plot are identified as linear or quadratic. The fully linear and quadratic models (6 and 4) utilised an identical surface discretisation to that shown. The results of these three analyses (4, 5 and 6) in which the material was not allowed to yield, are plotted within the GE/LSU strain gauge results in Fig. 6. Both the fully quadratic and mixed analyses are in excellent agreement with the strain gauge data. The difference between the two analyses is far less than the normal scatter in strain gauge data. The fully linear analysis, however, does not give either an accurate peak strain or a correct representation of the strain distribution near the notch.

After the verification of the elastic analysis a variety of inelastic analyses of the specimen were undertaken. Test specimen no. 7 was used for this

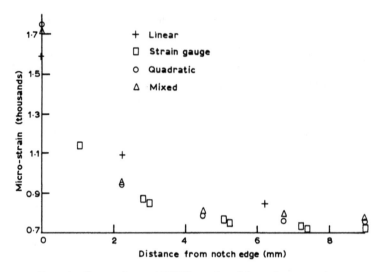

FIG. 6   Comparison of BEST results with strain gauge data.

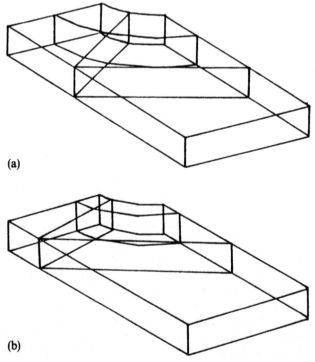

FIG. 7   Uniform and weighted mesh for inelastic analysis: (a) equally spaced mesh; (b) weighted mesh (20 surface patches, 40 boundary nodes, 4 volume cells).

purpose. The two-region mesh of Fig. 5 was slightly modified as shown in Fig. 7(a). Since plasticity develops only near the notch, volume modelling is required for only a small part of the inner ring around the notch. Accordingly, only four 20-noded volume cells were used around the cylindrical notch surface.

The first non-linear analysis carried out (using Von Mises model) for the specimen was a monotonic loading from zero load to the maximum load used in the test programme. The material was modelled as a Von Mises non-linearly strain-hardening material, with the equivalent stress versus equivalent plastic strain curve given by the equation (Domas *et al.*, 1982):

$$\varepsilon_{eq}^{p} = (\sigma_{eq}/210)^{1/0\cdot063} \tag{57}$$

where σ_{eq} is expressed in ksi.

The remaining material parameters for this non-linear analysis were $E = 23\cdot8 \times 10^3$ ksi, $v = 0\cdot3$ and $\sigma_0 = 100$ ksi. It was observed that the results of the present analysis gave the overall load versus the notch strain response too stiff. A careful examination of the experimental data for this specimen indicated that the maximum elastic strain response of the present mesh as well as that of model 5 (Table 1) was too low (1730 against 1820 from experiment). Instead of introducing more elements it was decided to make the surface elements near the notch surface smaller. The analysis was repeated using this weighted surface model (shown in Fig. 7(b)) to capture better the higher gradient near the notch. As shown in Fig. 8, the agreement

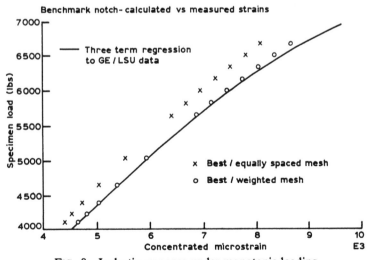

FIG. 8   Inelastic response under monotonic loading.

with the experimental data was excellent. The volume cells required no alteration even though the surface model was changed.

It should be noted that current engineering experience of the required discretisation for a problem is essentially based on two decades of finite element analysis. It is therefore instructive to compare the use of finite element modelling for this problem. The general purpose system MHOST (MARC-HOST) code (which used the 8-noded isoparametric solid element with a mixed variational principle together with the Loubignac iteration method to improve accuracy) was utilised for this purpose. Figure 9 shows

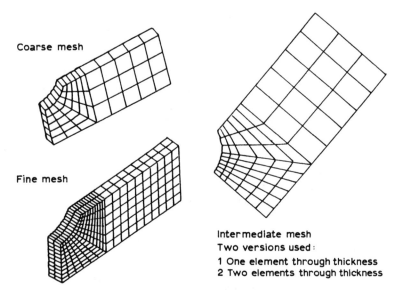

FIG. 9   Finite-element meshes for the benchmark notch specimen.

three equivalent finite element models (coarse, fine and intermediate) for this problem. Two versions of the intermediate mesh with one and two elements through the thickness were used. Figure 10 shows the results of all of these finite element analyses compared with the experimental data. It can be seen that apart from the coarse mesh all other finite element models agree with the data. The computing time for the coarse mesh was somewhat lower than that for the BEM analysis described above. All other finite element analyses were more expensive. However, in recent years the computing costs have become relatively insignificant in relation to the data

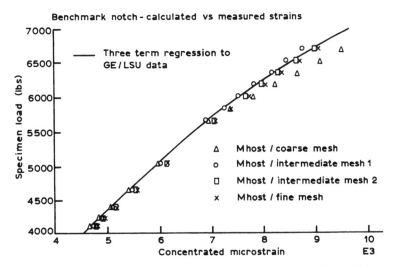

FIG. 10   Comparison between the finite-element results and experimental data for the benchmark notch specimen under monotonic loading.

preparation effort and it is obvious that the BEM models require a very small amount of input data.

For the cyclic loading analysis, both the inner yield surface (the loading surface) and the outer yield surface (the bounding surface) were represented by Von Mises yield surfaces (Banerjee *et al.*, 1987). The inelastic material parameters were chosen such that the model response under monotonic loading matches eqn (16). The material parameters were thus determined to be (Banerjee *et al.*, 1987)

$$h = 8 \cdot 2 \times 10^2 \text{ ksi}$$
$$\beta \text{ (the initial value)} = 100 \text{ ksi}$$
$$\sigma_{\text{ref}} = 210 \text{ ksi}$$
$$n = 13 \cdot 2$$

The weighted mesh shown in Fig. 7 was used in the BEM analysis by applying load increments of 5% and 20% of the specimen load at first yield. It is interesting to note that both results are almost identical and overall the agreement between the computed and experimental data (Fig. 11) is excellent. During the initial loading the size of the inner yield surface determines β at the first yield. Thereafter during each successive inelastic loading increment the inner yield surface moves in the stress space and the

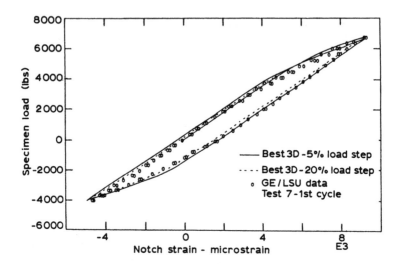

FIG. 11   Inelastic response under cyclic loading.

position of its centre needs to be stored each time the material becomes elastic as a result of unloading.

1.4.3 Turbine Component Analysis

In order to evaluate the capabilities of BEST3D for the analysis of real components, an analysis of a commercial cooled turbine blade geometry was carried out. It is expected that the use of this analysis as a benchmark problem will be continued as the capabilities of the present implementation improve for low cycle thermo-plastic calculations. This chapter discusses results for the elastic analysis of the blade.

The blade analysed is a cooled high turbine blade presently in service, where it is normally subject to mechanical loads (primarily centrifugal) and thermal loads. Of particular interest for this blade is the location and magnitude of the peak stress under the platform.

A BEST3D (surface) model was built for this problem. A view of the model with hidden lines removed is shown in Fig. 12. The model consists of five subregions. The interfaces between subregions are generally perpendicular to the radial direction. The characteristics of the model are summarised in Table 4. The system of equations for a fully linear analysis contains 1248 equations.

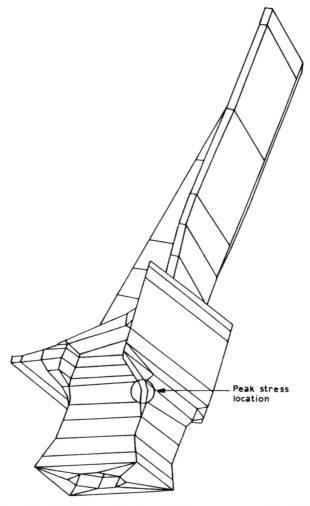

Peak stress
location

FIG. 12   BEST3D model of cooled high-pressure turbine blade.

TABLE 4

Subregion	Elements	Linear (nodes)	Quadratic (nodes)
1	60	61	180
2	86	85	256
3	107	98	303
4	106	96	298
5	80	76	232
Total	439	416	1269

Initially a fully linear analysis under centrifugal load (9549 rpm) was carried out. The total centrifugal load at various spanwise stations on the blade was compared with the standard design calculations for the blade. The agreement between the two totally independent calculations was excellent. The total centrifugal load for the blade was within 2% of the design calculation.

Study of blade tip deflections, load distribution over the base of the blade neck and concentrated stresses indicated reasonable qualitative agreement with three-dimensional finite element results. The time required to execute the BEST3D analysis was only 15 CPU min (on an IBM 3081) compared to 45 CPU min for the finite element (MARC) analysis.

While the use of stress contour plots demonstrated that the peak stress in the BEST3D analysis occurred in the correct location (under the trailing edge rail on the concave side of the blade), the value of the peak stress was too low (146 ksi vs 169 ksi). Improving this result using a fully quadratic analysis would have involved over 3700 degrees of freedom and required 45–60 min of computing time.

In order to improve local stress accuracy in the critical location while retaining computational efficiency, the mixed variation capability of BEST3D was employed. Quadratic variation was used over 16 elements (in only two subregions) in the immediate vicinity of the critical rail. The extent of the quadratic variation is shown in Fig. 13. The problem size increased by only 6% (from 1248 to 1317 degrees of freedom) and the computer time by

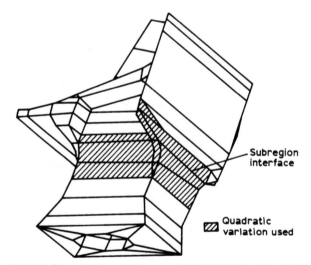

FIG. 13   Extent of quadratic variation used in BEST3D turbine blade analysis.

only 10% (to 16·5 CPU min). The peak stress increased to 174 ksi (on the surface), in excellent agreement with the MARC result of 169 ksi (at a slightly subsurface integration point).

The results of this study (one of the largest BEM analyses to date) clearly demonstrate both the capability of BEST3D for the analysis of real components and the power of the mixed variation boundary element method formulation.

1.4.4 Free-vibration of Twisted Plate

A complete set of BEST3D analyses has been done for the short, thick plate specimen used in the NASA-sponsored Joint Research Effort on Vibrations of Twisted Plates. The plate modelled measured 2 in (5 cm) in both the spanwise and chordwise directions and was 0·4 in (1 cm) thick. The BEST3D model utilised two subregions, each with one element through the specimen thickness and two elements in both the spanwise and chordwise directions. The meshes used for all five twist angles (0–60° in 15° increments) were topologically identical. The model used for the 60° plate is shown in Fig. 14. The plate was completely fixed at one end, as indicated in the figure.

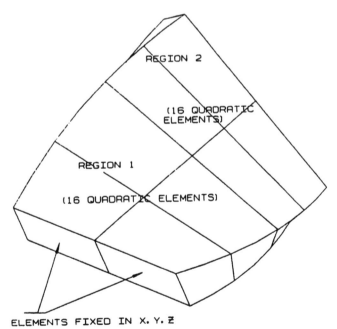

ELEMENTS FIXED IN X. Y. Z

FIG. 14   BEST3D model of 60° twisted plate.

This is the same boundary condition that was used in all of the finite element analyses in the study mentioned above.

The BEST3D analysis determined the first eight modes, for all five twist angles, within the 20 iterations normally allowed in the eigenvalue extraction routine. The BEST3D results have been compared with the four sets of finite element results which should be best suited for this problem. The finite element analyses considered were (Kielb *et al.*, 1985):

(1) Method L, using parabolic conoidal shell elements.
(2) Method M, using quadrilateral thick plate elements.
(3) Method N, using eight node isoparametric three-dimensional elements.
(4) Method O, using 16 node isoparametric three-dimensional elements.

The results for only the first bending and torsion modes are shown in Figs 15 and 16. Experimental results are also shown for comparison. In all these plots, the legend FE denotes finite element methods, BE denotes the BEST3D results and EX identifies the experimental results, while integers identify mode numbers. L, M, N or O is added to the FE identification to refer to the four methods listed above. A and B in the EX identification refer to the two different test sites (NASA-LeRC and the Air Force Aero Propulsion Laboratory).

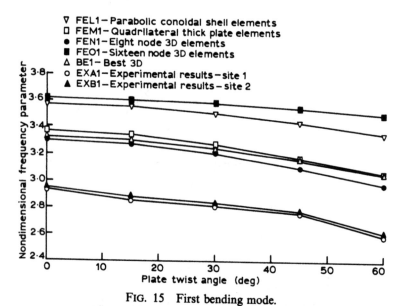

FIG. 15   First bending mode.

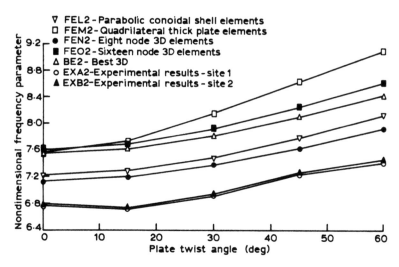

FIG. 16   First torsion mode.

As Figs 15 and 16 indicate, there is considerable scatter among the finite element results, especially for the higher modes. The BEST3D results clearly fit within the range of the finite element results. BEST3D shows the closest agreement with the finite element results obtained using thick plate elements (method M) and 16 node isoparametric elements (method O). The only analytical results for comparison are natural frequencies for first bending and first torsion for the flat (untwisted) plate. BEST3D and the thick plate finite element method (method M) show closer and more consistent agreement with these analytical results than do any of the other results.

The experimental results are consistently lower than the analytical values for all modes. This is particularly exaggerated for the first edgewise bending mode (not shown here), especially at low twist angles. The lower experimental values are thought to be due to the compliance of the specimen base and fixturing. For the first two modes, both BEST3D and finite element method M correctly predict the qualitative variation of natural frequency response with twist angle, although method M predicts a larger variation than was observed for the first torsion mode.

Based on results presently available, it seems clear that BEST3D predicts the plate response at least as consistently as do any of the finite element models.

In order to capture the effects of base compliances, the twisted plate

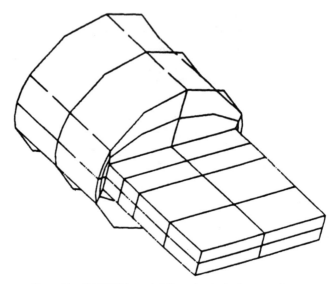

FIG. 17   BEST3D model (3-region) of plate specimen.

geometries were then analysed using a model which includes the specimen base (Fig. 17). The results, at all twist angles, for both the first mode (easy bending) and the second mode (torsion) were dramatically improved relative to the model without a base (Figs 18 and 19) and show excellent agreement with the experimental results. Figures 20 and 21 show the corresponding mode shapes for a typical blade.

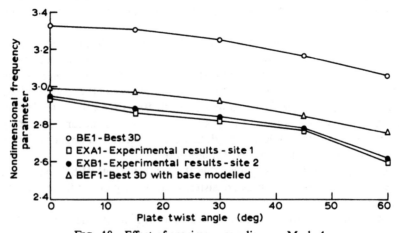

FIG. 18   Effect of specimen compliance—Mode 1.

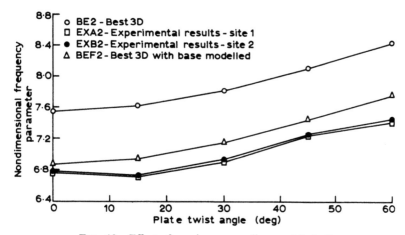

FIG. 19 Effect of specimen compliance—Mode 2.

FIG. 20 First bending mode shape.

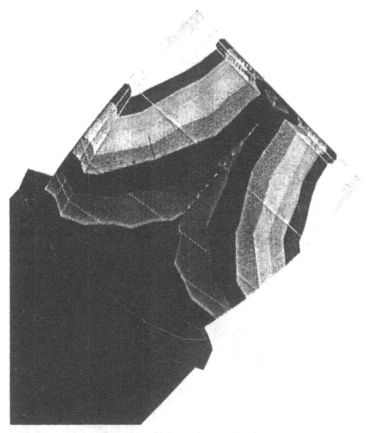

FIG. 21   First torsion mode shape.

1.5 CONCLUSIONS

The examples presented in this chapter clearly demonstrate that many of
the potentials of BEM as a superior stress analysis tool are now slowly
being realised. The general purpose code BEST3D developed under the
NASA inelastic methods contract has the potential to analyse satisfactorily
many of the gas turbine engine structures accurately and efficiently. The
anisotropic component of the code which is developed from the work of
Wilson and Cruse (1978) is now undergoing extensive verification and will
be reported shortly.

ACKNOWLEDGEMENTS

It should be emphasised that a large undertaking such as the development of BEST3D is possible only through the collective efforts of a large number of people. To Dr Chris Chamis of NASA Lewis Center, Cleveland, who had the vision to conceive this project, and to Dr Edward Todd of Pratt & Whitney for inspired leadership we express our sincere gratitude. A very significant contribution to the development effort was made by David Snow of Pratt & Whitney, who had the major responsibility for the validation examples.

To Drs T.G. Davies and G.D. Manolis of the Department of Civil Engineering, State University of New York at Buffalo, who carried out the early part of the investigations for the inelastic and dynamic analyses, respectively, we express our deepest appreciation for their work. A number of former and present graduate students (at SUNY, Buffalo) of exceptional academic ability and programming skill contributed substantially to the development of BEST3D. In particular, we are indebted to Drs S.T. Raveendra and D. Henry for their major contributions in inelastic analyses, to Dr N.B. Yousif for her contributions in inelastic constitutive equations, to Dr R. Sen for the early part of the work on the transient dynamic analyses, to Dr S. Ahmad and Dr Wang for their major contributions in periodic dynamic, transient time domain dynamic analysis algorithms, and to Dr G. Dargush for his major contributions in thermal problems.

REFERENCES

AHMAD, S. & BANERJEE, P.K. (1986). Free vibration analysis by BEM using particular integrals, *J. Eng. Mech. Div., Am. Soc. Civ. Eng.*, **112**(7), 682–95.

BANERJEE, P.K. & BUTTERFIELD, R. (1975). Boundary element methods in geomechanics, Chapter 16 in: *Finite Element in Geomechanics*, Ed. G. Gudehus, John Wiley, Chichester, UK; *Proc. NMSRM*, University of Karlsruhe, FRG.

BANERJEE, P.K. & BUTTERFIELD, R. (1981). *Boundary Element Methods in Engineering Science*, McGraw-Hill, London. Also US Edition, New York, 1983; Russian Edition, Moscow, 1984.

BANERJEE P.K. & CATHIE, D.N. (1980). A direct formulation and numerical implementation of the boundary element method for two-dimensional problems of elastoplasticity, *Int. J. Mech. Sci.*, **22**, 233–45.

BANERJEE, P.K. & DAVIES, T.G. (1984). Advanced implementation of boundary element methods for three-dimensional problems of elastoplasticity and viscoplasticity. Chapter 1 in: *Developments in Boundary Element Methods—3*, Eds P.K. Banerjee and S. Mukherjee, Elsevier Applied Science Publishers, London, 1–26.

BANERJEE, P.K. & RAVEENDRA, S.T. (1986). Advanced boundary element analysis of two- and three-dimensional problems of elastoplasticity, *Int. J. Numer. Methods Eng.*, **23**, 985–1002.

BANERJEE, P.K. & RAVEENDRA, S.T. (1987). A new boundary element formulation for two-dimensional elastoplastic analysis, *J. Eng. Mech. Div., Am. Soc. Civ. Eng.*, **113**(2), 252–65.

BANERJEE, P.K., CATHIE, D.N. & DAVIES, T.G. (1979). Two- and three-dimensional problems of elastoplasticity. Chapter 4 in: *Developments in Boundary Element Methods—1*, Eds P.K. Banerjee and R. Butterfield, Applied Science Publishers, London, 65–95.

BANERJEE, P.K., WILSON, R.B. & MILLER, N.M. (1985). Development of a large BEM system for three-dimensional inelastic analysis, *Proc. ASME Conf. on Advanced Topics in Boundary Element Analysis*, Eds T.A. Cruse, A.B. Pifko and H. Armen, AMD Vol. 72, ASME, New York, 1–20.

BANERJEE, P.K., WILSON, R.B. & RAVEENDRA, S.T. (1987). Advanced applications of BEM to three-dimensional problems of monotonic and cyclic plasticity, *Int. J. Mech. Sci.* **29**(9), 637–53.

BANERJEE, P.K., WILSON, R.B. & MILLER, N.M. (1988a). Advanced elastic and inelastic three-dimensional analysis of gas turbine engine structures by BEM, *Int. J. Numer. Methods Eng.*, **26**, 393–411.

BANERJEE, P.K., AHMAD, S. & WANG, H.C. (1988b). A new BEM formulation for acoustic eigenfrequency analysis, *Int. J. Numer. Methods Eng.*, **26**, 1299–309.

DOMAS P.A., SHARPE, W.N., WARD, M. & YAU, J. (1982). *Benchmark Notch Test for Life Prediction*, NASA CR-165571.

HENRY, D.P., JR, & BANERJEE, P.K. (1987). A thermoplastic BEM analysis for substructured axisymmetric bodies, *J. Eng. Mech. Div., Am. Soc. Civ. Eng.*, **113**(12), 1880–900.

HENRY, D.P., JR, PAPE, D.A. & BANERJEE, P.K. (1987). A new axisymmetric BEM formulation for body forces using particular integrals, *J. Eng. Mech. Div., Am. Soc. Civ. Eng.*, **113**(5), 671–88.

KIELB, R.E., LEISSA, A.W., MACBAIN, J.C. & CARNEY, K.S. (1985). *Joint Research Effort on Vibration of Twisted Plates—Phase I, Final Results*, NASA Reference Publication 1150.

LACHAT, J.C. & WATSON, J.O. (1976). Effective numerical treatment of boundary integral equations: a formulation for three-dimensional elastostatics, *Int. J. Numer. Methods Eng.*, **10**, 991–1005.

MANOLIS, G.D. & BANERJEE, P.K. (1986). Conforming versus non-conforming boundary elements in three-dimensional elastostatics, *Int. J. Numer. Methods Eng.*, **23**, 1885–904.

NARDINI, D. & BREBBIA, C.A. (1982). A new approach to free-vibration analysis using boundary elements, *Proc. 4th Conf. on BEM*, University of Southampton, UK, 313–26.

PAPE, D.A. & BANERJEE, P.K. (1987). Treatment of body forces in 2-D elastostatic BEM using particular integrals, *J. Appl. Mech. ASME*, **54**, 866–71.

RAVEENDRA, S.T. (1984). Advanced development of BEM for two- and three-dimensional nonlinear analysis, Ph.D. Thesis, State University of New York at Buffalo.

SAAD, Y. (1984). Chebyshev acceleration technique for solving nonsymmetric eigenvalue problems, *Math. Comp.*, **42**(166), 567–88.

WANG, H.C. & BANERJEE, P.K. (1988). Axisymmetric free-vibration problems by the boundary element method, *J. Appl. Mech., ASME*, **55**, 437–42.

WILSON, R.B. & BANERJEE, P.K. (1986). *3-D Inelastic Analysis Methods for Hot Section Components*, Third Annual Status Report, Vol. II, NASA Contract Report NAS 179517.

WILSON, R.B. & CRUSE, T.A. (1978). Efficient implementation of anisotropic three-dimensional boundary integral equation stress analysis, *Int. J. Numer. Methods Eng.*, **12**, 1383–97.

WILSON, R.B., SNOW, D.W. & BANERJEE, P.K. (1985). Stress analysis of gas turbine engine structures using boundary element methods, *Proc. ASME Conf. on Advanced Topics in Boundary Element Analysis*, Eds T.A. Cruse, A.B. Pifko and H. Armen, AMD Vol. 72, ASME, New York, 45–64.

Chapter 2

ADVANCED APPLICATIONS OF BEM
TO INELASTIC ANALYSIS OF SOLIDS

P.K. BANERJEE and D.P. HENRY, JR

Department of Civil Engineering, State University of New York at Buffalo, USA

SUMMARY

A number of recently developed boundary element solution algorithms are used for solving a range of elastoplastic and thermoplastic problems. The first algorithm is the conventional iterative procedure in which the unknown boundary solution and the initial stress (or strain) rates are found together in an incremental iterative fashion. The second algorithm is a new variable stiffness type approach in which the incremental boundary solution is obtained in a direct (non-iterative) manner. The third procedure, which is entirely new, differs from the previous two in that volume integration is not required to incorporate the non-linear effects in the analysis. Instead, initial stress rates are introduced in the boundary element system via particular integrals.

All these approaches are presented in a general manner for axisymmetric, two- and three-dimensional analyses. The formulation is implemented in a general purpose, multi-region system that utilises quadratic isoparametric shape functions to model the geometry and field variables of the body and can admit up to 15 substructured regions of different material properties.

2.1 INTRODUCTION

Starting with the elastoplastic boundary element formulation introduced by Swedlow and Cruse (1971), the first two-dimensional elastoplastic analysis was developed by Riccardella (1973), the first viscoplastic analyses by Kumar and Mukherjee (1975) and Chaudonneret (1977), the first three-dimensional analyses by Banerjee et al. (1979), and the first axisymmetric analysis by Cathie and Banerjee (1980). Since then these formulations have been improved by many researchers. These early elastoplastic BEM formulations were all based on iterative procedures which work successfully, but often take an unduly large number of iterations to converge to the correct solution, particularly in problems involving a high degree of non-linearity such as the loading close to the collapse state of stress when a significant amount of plastic zones develop. For this reason, Banerjee and Raveendra (1986) developed an advanced implementation of the iterative algorithm (Banerjee et al., 1979; Banerjee and Davies, 1984) which has a time saving feature which reduces the number of iterations needed for convergence by utilising the past history of initial stress rates to estimate the value of the initial stress rates for the next load increment. The method was found to reduce substantially the time needed for iteration and this procedure has been adopted in the iterative results presented in this chapter. Nevertheless, the iterative procedure still often has trouble converging to a correct solution for a crude mesh, particularly close to the collapse state of stress. It is to this end that Raveendra (1984) and Banerjee and Raveendra (1987) presented the first direct or 'non-iterative', two-dimensional elastoplastic analysis which is comparable to the variable stiffness method in finite element analysis. The method proved successful and has recently been extended to axisymmetric analysis by Henry and Banerjee (1988a) and to the three-dimensional case by Banerjee and Henry (1987) and Banerjee et al. (1988a).

The need for only surface discretisation is a significant advantage BEM has over other methods requiring full domain discretisation. However, this advantage is partially diminished in an elastoplastic analysis when volume integration is required. In the latter part of this chapter, a novel approach is introduced in which non-linear effects are incorporated in the boundary element system without the need for volume integration. The method is based on the well-known concept of developing a solution of an inhomogeneous differential equation by means of a complementary function and a particular integral, and has been developed recently by Henry (1987) and Henry and Banerjee (1988b, c).

The use of particular integrals in BEM was tentatively discussed by Watson (1979) and by Banerjee and Butterfield (1981). Ahmad and Banerjee (1986) successfully employed the concept in a two-dimensional free vibration analysis, followed by Wang and Banerjee (1988) who developed the axisymmetric counterpart. The theory was then modified by Banerjee, Ahmad and Wang (Banerjee *et al.*, 1988*a*) for acoustic eigenfrequency analysis. Concurrent to this development, Henry and Banerjee (1988*b*) formulated a steady-state and transient, uncoupled thermoelasticity which employed particular integrals for incorporating thermal body forces in the analysis, which was extended to elastoplastic analysis by Henry and Banerjee (1988*c*). Particular integral formulations have also been presented for gravitational and centrifugal body forces in axisymmetric, two- and three-dimensional stress analyses (Pape and Banerjee, 1987; Henry *et al.*, 1987; Banerjee *et al.*, 1988*b*).

The first half of this chapter is dedicated to the presentation of the newly developed variable stiffness method for the substructured, elastoplastic analysis of axisymmetric, two- and three-dimensional media with (variable) strain hardening. In the second part of the chapter the newly developed method of particular integrals is described. All these methods represent realistic practical examples, including, for the first time, three-dimensional problems analysed using the variable stiffness method. To the best of the authors' knowledge, a comparable boundary element analysis for multi-region or substructured inelastic two and three-dimensional and axisymmetric problems has not been attempted before.

2.2 ELASTOPLASTIC FLOW EQUATIONS

For a standard elastoplastic flow problem the evolution equation for plastic flow is governed by

$$F(\sigma_{ij}, h) = 0$$

and

$$\dot{\varepsilon}_{ij}^{\mathrm{p}} = \lambda \frac{\partial F}{\partial \sigma_{ij}} \tag{1}$$

These equations together with the consistency relations (i.e. the stress point must remain on a newly developing yield surface characterised by a change in the hardening parameter h) lead to the following expression for the unknown plastic flow factor λ:

$$\dot{\lambda} = L_{ij}^{\sigma} \dot{\sigma}_{ij} \tag{2}$$

where

$$L^{\sigma}_{ij} = \frac{1}{H} \frac{\partial F}{\partial \sigma_{ij}}$$

$$H = -\left(\frac{\partial F}{\partial \varepsilon^{p}_{mn}} + \frac{\partial F}{\partial h} \frac{\partial h}{\partial \varepsilon^{p}_{mn}} \right) \frac{\partial F}{\partial \sigma_{mn}}$$

It should be noted that L^{σ}_{ij} depends upon the current state variables, not on the incremental quantities.

However, the relationship given by eqn (2) does not exist for ideal plasticity as H vanishes for zero hardening. This can be avoided by reformulating the above expression in terms of strain increments:

$$\dot{\lambda} = L^{\varepsilon}_{ij} \dot{\varepsilon}_{ij} \qquad (3)$$

where

$$L^{\varepsilon}_{ij} = \frac{1}{H'} \frac{\partial F}{\partial \sigma_{kl}} D^{e}_{klij}$$

$$H' = \frac{\partial F}{\partial \sigma_{kl}} D^{e}_{klmn} \frac{\partial F}{\partial \sigma_{mn}} - \left(\frac{\partial F}{\partial \varepsilon^{p}_{kl}} + \frac{\partial F}{\partial h} \frac{\partial h}{\partial \varepsilon^{p}_{kl}} \right) \frac{\partial F}{\partial \sigma_{kl}}$$

and D^{e}_{ijkl} is the elastic constitutive tensor. It is evident that H' does not vanish for zero hardening (ideal plasticity).

For a Von Mises material we have

$$F = \sqrt{\tfrac{3}{2} S_{ij} S_{ij}} - h\sqrt{\tfrac{2}{3} \varepsilon^{p}_{ij} \varepsilon^{p}_{ij}} \qquad (4)$$

where

$$S_{ij} = \sigma_{ij} - \tfrac{1}{3} \delta_{ij} \sigma_{kk}$$

h = the slope of the uniaxial equivalent stress–plastic strain curve.

Equation (4) can be used to obtain explicit forms of L^{σ}_{ij}, L^{ε}_{ij} or the incremental stress–strain relations in the usual manner.

2.3 THE INTEGRAL FORMULATION FOR ELASTOPLASTICITY

2.3.1 Volume Integral Formulations

The conventional boundary integral equation for displacement satisfying the governing equation for plasticity can be expressed (Banerjee and

Butterfield, 1981; Banerjee and Raveendra, 1986) as:

$$C_{ij}(\xi)\dot{u}_i(\xi) = \int_S [G_{ij}(x, \xi)\dot{t}_i(x) - F_{ij}(x, \xi)\dot{u}_i(x)] \, dS(x)$$

$$+ \int_V B_{ikj}(x, \xi)\dot{\sigma}^o_{ik}(x) \, dV(x) \tag{5}$$

where G_{ij}, F_{ij} and B_{ijk} are kernel functions. The two- and three-dimensional kernel functions are given by Banerjee and Raveendra (1986) and the axisymmetric kernels can be found in Henry and Banerjee (1987, 1988a). Because these are lengthy algebraic expressions, they are not reproduced here.

The initial stress rate σ^o_{ij} can be defined in terms of the plastic strain rate as:

$$\dot{\sigma}^o_{ij} = D^\varepsilon_{ijkl}\dot{\varepsilon}^p_{kl} \tag{6}$$

Body forces, if present, can be incorporated via a particular integral (Banerjee et al., 1989; Henry et al., 1987; Pape and Banerjee, 1987).

The integral equation for the strain at an interior point is found analytically by substituting eqn (5) (with $C_{ij} = \delta_{ij}$) into the strain–displacement relations, where differentiation is with respect to the field point ξ:

$$\dot{\varepsilon}_{ij}(\xi) = \int_S [G^\varepsilon_{kij}(x, \xi)\dot{t}_k(x) - F^\varepsilon_{kij}(x, \xi)\dot{u}_k(x)] \, dS(x)$$

$$+ \int_V B^\varepsilon_{klij}(x, \xi)\dot{\sigma}^o_{kl}(x) \, dV(x) + J^\varepsilon_{klij}\dot{\sigma}^o_{kl}(\zeta) \tag{7}$$

By introducing this result into the inelastic stress–strain equation ($\dot{\sigma}_{ij} = D_{ijkl}\dot{\varepsilon}_{kl} - \dot{\sigma}^o_{ij}$), the stress integral equation is derived:

$$\dot{\sigma}_{ij}(\xi) = \int_S [G^\sigma_{kij}(x, \xi)\dot{t}_k(x) - F^\sigma_{kij}(x, \xi)\dot{u}_k] \, dS(x)$$

$$+ \int_V B^\sigma_{klij}(x, \xi)\dot{\sigma}^o_{kl}(x) \, dV(x) + J^\sigma_{klij}\dot{\sigma}^o_{kl}(\xi) \tag{8}$$

The last integral in eqns (7) and (8) is strongly singular and must be treated in the Cauchy-principle sense with additional jump terms J^σ and J^ε (Banerjee and Raveendra, 1986). All kernel functions G, F and B as well as the jump terms are defined in Banerjee and Raveendra (1986) and Henry and Banerjee (1987, 1988a).

For a point on the boundary, all kernel functions of the stress rate equation are all strongly singular and difficult to integrate numerically. However, the stress rate on the boundary can be obtained from the boundary traction and displacement rates without any integration by substitution of the inelastic stress–strain relation in place of Hooke's Law (Banerjee and Raveendra, 1986; Henry and Banerjee, 1987). In this procedure the stress–strain relations, the Cauchy traction equation and the equations relating local and global gradients of displacement are utilised to write an expression for the bounday stress rates. For numerical purposes, this equation can be arranged in a form consistent with the discretised form of eqn (7) or (8) (Banerjee and Raveendra, 1986).

2.3.2 Variable Stiffness Plasticity Formulation

The above non-linear formulations include initial stress rates in the governing equations which are not known *a priori*, and therefore are conventionally solved using an iterative procedure. A non-iterative, direct solution method is made feasible by reducing the number of unknowns in the governing equations by utilising certain features of the incremental theory of plasticity expressed by eqns (1), (2) and (3).

The initial stress rates σ_{ij}^0 appearing in the incremental form of integral equations (5) to (8) can be expressed in the context of an elastoplastic deformation as:

$$\dot{\sigma}_{ij}^0 = K_{ij}\dot{\lambda} \tag{9}$$

where

$$K_{ij} = D_{ijkl}^e \frac{\partial F}{\partial \sigma_{kl}}$$

Substituting eqns (2) and (9) in eqns (5) and (8) yields (in the absence of body forces):

$$C_{ij}(\xi)\dot{u}_i(\xi) = \int_S [G_{ij}(x,\xi)\dot{t}_i(x) - F_{ij}(x,\xi)\dot{u}_i(x)]\,dS(x)$$

$$+ \int_V B_{ipj}(x,\xi)K_{ip}(x)\dot{\lambda}(x)\,dV(x) \qquad (10)$$

$$\dot{\lambda}(\xi) = L^\sigma_{jk}(\xi)\int_S [G^\sigma_{ijk}(x,\xi)\dot{t}_i(x) - F^\sigma_{ijk}(x,\xi)\dot{u}_i(x)]\,dS(x)$$

$$+ L^\sigma_{jk}(\xi)\left[\int_V B^\sigma_{ipjk}(x,\xi)K_{ip}(x)\dot{\lambda}(x)\,dV(x) + J^\sigma_{ipjk}K_{ip}(x)\dot{\lambda}(x)\right] \quad (11)$$

Equations (10) and (11) can be solved simultaneously to evaluate the unknown values of displacements, traction rates and the scalar variable $\dot{\lambda}$.

Although eqn (11) can be applied to any elastoplastic strain hardening problem, L^σ_{ij} becomes indeterminate for the case of ideal plasticity (zero strain hardening) and therefore cannot be used in this circumstance. This minor problem, however, can be circumvented either by specifying a small amount of strain hardening or more appropriately by replacing eqn (11) by the strain rate equations (7) and (3). Thus using eqns (3) and (9) in eqn (7) we obtain

$$\dot{\lambda}(\xi) = L^\varepsilon_{jk}(\xi)\int_S [G^\varepsilon_{ijk}(x,\xi)\dot{t}_i(x) - F^\varepsilon_{ijk}(x,\xi)\dot{u}_i(x)]\,dS(x)$$

$$+ L^\varepsilon_{jk}(\xi)\left[\int_V B^\varepsilon_{ipjk}(x,\xi)K_{ip}(x)\dot{\lambda}(x)\,dV(x) + J^\varepsilon_{ipjk}K_{ip}(x)\dot{\lambda}(x)\right] \quad (12)$$

Equation (12) can now be applied to any problems of elastoplasticity (both strain hardening and ideal plasticity).

2.3.3 Numerical Implementation
The numerical implementation of BEM has received much attention in recent years, and therefore it will be only briefly discussed here. The interested reader should refer to Watson (1979), Raveendra (1984), Banerjee and Raveendra (1986), Henry (1987), and Henry and Banerjee (1987).

The body under consideration and eqns (10) and (11) or (12) that are used to represent it, are divided into boundary elements and volume cells. The boundary elements cover the entire boundary surface and the cells are located in the domain where plastic yielding is expected. Quadratic isoparametric curvilinear shape functions are used to approximate the geometry and the field variables across the boundary elements and over volume cells in terms of their nodal values. The integrands of the discretised equations are evaluated using self-adaptive standard Gaussian quadrature techniques. It is important to note that the functions $L_{ij}(\xi)$ and $K_{ij}(x)\dot{\lambda}(x)$ are expressed as nodal quantities via shape functions, so that these quantities can be brought outside the integration of the discretised integrals of eqns (10) to (12). Therefore, the integration need only be performed once.

2.3.4 Initial Stress Expansion Technique
The B^σ_{ijkl} and B^ε_{ijkl} kernels are strongly singular when integrated over a volume containing the field point, and accurate numerical integration at this singular point for the most general case is costly and difficult (Banerjee and Raveendra, 1986). The difficulties, however, can be alleviated by determining the coefficient of the singular node in an indirect manner (Henry and Banerjee, 1987, 1988a; Banerjee et al., 1989).

In this procedure, the coefficients of the stress equations related to the non-singular nodes are integrated in the usual manner using Gaussian integration with appropriate element subdivisions. In each stress equation there remain three undetermined coefficients in two dimensions, or six in three dimensions, or four in axisymmetry, corresponding to the initial stress at the singular node, which cannot be evaluated easily. In a manner analogous to the 'rigid body' motion technique for the surface equations, each of these coefficients is calculated by assuming one (each row) of the three (six or four) admissible initial stress states and compatible displacement fields given in Tables 1, 2 and 3. In an unrestrained body the resulting stresses are either zero (two and three dimensions) or can be determined (axisymmetry) for an assumed initial stress field. Thus, for each assumed initial stress state, one unknown coefficient in each stress equation can be calculated by using all the other coefficients which are non-singular. It should be noted that in order to apply this method the entire region must be covered with cells. In an isothermal analysis in which the elastic yield zone is small, this technique appears inefficient since it requires the presence of volume cells throughout the elastic region which otherwise would be unnecessary. Although this is true in a single region programme, such is not the case in a multi-region code, since the technique is applied to each region

TABLE 1
STRESS STATES FOR INITIAL STRESS EXPANSION TECHNIQUE IN TWO-DIMENSIONAL PLANE STRAIN (PLANE STRESS) ANALYSIS

Stress state	Coefficient to be determined corresponds to	Nodal values of assumed stress state[a]				
		σ_{xx}^o	σ_{yy}^o	σ_{xy}^o	u_x	u_y
1	σ_{xx}^o	E	0	0	$(1-v^2)x$	$-v(1+v)y$
2	σ_{yy}^o	0	E	0	$-v(1+v)x$	$(1-v^2)y$
3	σ_{xy}^o	0	0	E	$(1+v)y$	$(1+v)x$

[a] E = modulus of elasticity; v = Poisson's ratio; x, y = nodal coordinates. All stresses and tractions are zero for all stress states.

The stress states for two-dimensional plane strain analysis given here can be applied to the plane stress case, if the following modified material parameters are used: $\bar{E} = E(1+2v)/(1+v)^2$; $\bar{v} = v/(1+v)$.

TABLE 2
STRESS STATES FOR INITIAL STRESS EXPANSION TECHNIQUE IN THREE-DIMENSIONAL ANALYSIS

Stress state	Coefficient to be determined corresponds to	Nodal values of assumed stress state[a]								
		σ_{xx}^o	σ_{yy}^o	σ_{zz}^o	σ_{xy}^o	σ_{xz}^o	σ_{yz}^o	u_x	u_y	u_z
1	σ_{xx}^o	E	0	0	0	0	0	x	$-vy$	$-vz$
2	σ_{yy}^o	0	E	0	0	0	0	$-vx$	y	$-vz$
3	σ_{zz}^o	0	0	E	0	0	0	$-vx$	$-vy$	z
4	σ_{xy}^o	0	0	0	E	0	0	$(1+v)y$	$(1+v)x$	0
5	σ_{xz}^o	0	0	0	0	E	0	$(1+v)z$	0	$(1+v)x$
6	σ_{yz}^o	0	0	0	0	0	E	0	$(1+v)z$	$(1+v)y$

[a] E = modulus of elasticity; v = Poisson's ratio; x, y, z = nodal coordinates. All stresses and tractions are zero for all stress states.

independently and only used in regions where volume cells exist, which happens to be the most efficient inelastic BEM analysis (Banerjee et al., 1987). Therefore, the zones in which plastic yielding is expected to occur are isolated in separate regions fully populated with cells, and the elastic regions remain free of any volume cells. Telles (1983) also discussed this idea of using simple initial strain fields for determining these coefficients for a two-dimensional elastoplastic problem.

TABLE 3

STRESS STATES FOR INITIAL STRESS EXPANSION TECHNIQUE IN AXISYMMETRIC ANALYSIS

Stress state	Coefficient to be determined corresponds to	Nodal values of assumed stress state[a]											
		σ_{rr}	σ_{zz}	$\sigma_{\theta\theta}$	σ_{rz}	σ_{rr}^{o}	σ_{zz}^{o}	$\sigma_{\theta\theta}^{o}$	σ_{rz}^{o}	u_r	u_z	t_r	t_z
1	σ_{rr}^{o}	$\dfrac{(2-\nu)E}{3(1-\nu)C}r$	$\dfrac{\nu E}{(1-\nu)C}r$	$2\sigma_{rr}$	0	0	0	$-\dfrac{E}{C}r$	0	$\dfrac{(1+\nu)(1-2\nu)}{3(1-\nu)C}r^2$	0	$\sigma_{rr}n_r$	$\sigma_{zz}n_z$
2	σ_{zz}^{o}	0	0	0	0	E	0	E	0	$(1-\nu)r$	$-2\nu z$	0	0
3	σ_{zz}^{o}	0	0	0	0	0	E	0	0	$-\nu r$	z	0	0
4	σ_{rz}^{o}	0	0	0	0	0	0	0	μ	0	r	0	0

[a] E = modulus of elasticity; μ = shear modulus; ν = Poisson's ratio; r, z = nodal coordinates; n_r, n_z = normals of the boundary; C = arbitrary parameter with dimensions of length, added to ensure dimensional homogeneity. The value can be equated to one for simplicity.

2.3.5 System Equations for the Variable Stiffness Elastoplasticity

The displacement and stress rate equations are assembled by collecting the known and unknown values of traction and displacement rate and their coefficients together. The assembled equations are modified appropriately for cases where the functions (boundary conditions) are referred to local coordinates. The equilibrium and compatibility conditions are invoked at common interfaces in a manner described by Banerjee and Butterfield (1981) and the final system equations can be cast as (Raveendra, 1984; Banerjee and Raveendra, 1987):

$$\mathbf{A}^b \dot{\mathbf{x}} = \mathbf{B}^b \dot{\mathbf{y}} + \mathbf{C}^b \mathbf{K} \dot{\lambda} \tag{13a}$$

$$\dot{\lambda} = \mathbf{L} \mathbf{A}^\sigma \dot{\mathbf{x}} + \mathbf{L} \mathbf{B}^\sigma \dot{\mathbf{y}} + \mathbf{L} \mathbf{C}^\sigma \mathbf{K} \dot{\lambda} \tag{13b}$$

where

$\dot{\mathbf{y}}$ are the known incremental boundary conditions,

$\dot{\mathbf{x}}$ are the unknown,

$\mathbf{A}^b, \mathbf{B}^b, \mathbf{C}^b$ are the coefficient matrices of the boundary (displacement) system, and

$\mathbf{A}^\sigma, \mathbf{B}^\sigma, \mathbf{C}^\sigma$ are the coefficient matrices of the stress equations.

Matrices $\mathbf{A}^b$, $\mathbf{B}^b$, $\mathbf{C}^b$ and $\mathbf{A}^\sigma$, $\mathbf{B}^\sigma$, $\mathbf{C}^\sigma$ are constant for a given problem and are identical to those used in the iterative algorithm, whereas matrices $\mathbf{K}$ and $\mathbf{L}$ are dependent upon state variables, and are assumed to be constant during each small load step. Furthermore, in a substructured system the matrices $\mathbf{A}^b$ and $\mathbf{B}^b$ are block banded and, since each substructured region can only communicate through the common interfaces, matrices $\mathbf{C}^b$, $\mathbf{A}^\sigma$, $\mathbf{B}^\sigma$ and $\mathbf{C}^\sigma$ are block diagonal.

Note that only equations for the cell nodes that are expected to yield during the current load step are used in the system equation (13b). At elastic nodes, the values of λ are assumed to be zero. This renders a small set of equations (13b) at the start of the analysis which eventually grows larger as more nodes yield.

These equations can be rewritten as:

$$\mathbf{A}^b \dot{\mathbf{x}} = \dot{\mathbf{b}}^b + \mathbf{C}^{\lambda b} \dot{\lambda} \tag{14a}$$

$$\dot{\lambda} = \mathbf{A}^\lambda \dot{\mathbf{x}} + \dot{\mathbf{b}}^\lambda + \mathbf{C}^\lambda \dot{\lambda} \tag{14b}$$

Upon rearranging eqn (14b) we have

$$\mathbf{H} \dot{\lambda} = \mathbf{A}^\lambda \dot{\mathbf{x}} + \dot{\mathbf{b}}^\lambda \tag{15}$$

where

$$\dot{b}^b = B^b \dot{y}, \ C^{\lambda b} = C^b K, \ A^\lambda = LA^\sigma,$$
$$\dot{b}^\lambda = LB^\sigma \dot{y}, \ C^\lambda = LC^\sigma K, \ H = I - C^\lambda, \text{ and}$$

I is the identity matrix.

Equation (15) can then be recast as

$$\dot{\lambda} = A^\circ \dot{x} + \dot{b}^\circ \tag{16}$$

where

$$A^\circ = H^{-1} A^\lambda$$
$$\dot{b}^\circ = H^{-1} \dot{b}^\lambda$$

Substituting the above equation into eqn (14a) results in the final system equations:

$$A^r \dot{x} = \dot{b}^r \tag{17}$$

where

$$A^r = A^b - C^{\lambda b} A^\circ$$
$$\dot{b}^r = \dot{b}^b + C^{\lambda b} \dot{b}^\circ$$

Equation (17) is constructed and solved to evaluate the unknown vector $\dot{x}$ at boundary nodes for every increment of loading. This formulation is similar to the variable stiffness approach used in the finite element method since the system matrix on the boundary as well as the vector on the right-hand side are modified for each increment of loading. For a multi-region system the inversion of H can be carried out for each region separately. Note that the size of H is equal to the number of nodes that are assumed to be plastic for the current load increment.

2.3.6 Solution Process for Variable Stiffness Approach
As previously mentioned, the solution process does not involve any iterative procedure; instead the substantial part of the solution effort is spent on assembly of the system equations for each load step. These operations can be described as follows:

(a) Impose an arbitrary boundary loading and solve the elastic problem in the usual manner.
(b) Scale the elastic solution such that the most highly stressed node is at yield.

(c) Apply a small load increment (usually less than 5% of the yield load) and compute **K** and **L** matrices using the past stress history.

(d) Form the system eqn (17) and solve for $\dot{\mathbf{x}}$.

(e) Evaluate the initial stress rates $\dot{\sigma}^0$ using eqns (9) and (16):

$$\dot{\sigma}^0 = \mathbf{K}\dot{\lambda} = \mathbf{K}[\mathbf{A}^0\dot{\mathbf{x}} + \dot{\mathbf{b}}^0]$$

(f) Evaluate interior quantities, displacement and stress rates using the discretised form of eqns (5) and (8).

(g) Return to step (c) if the strains are less than a specified norm. Otherwise, failure is assumed to have occurred.

It is of importance to note that the matrices **K** and **L** do not exist in the elastic region, therefore the equations involving these matrices are formed (and σ^0 is determined) only for the nodes that are at yield. Furthermore, any small deviation from the yield surface can be corrected by applying the stress rate difference (i.e. the initial stress rate) during the next load step. Essentially a radial return algorithm is followed here.

2.4 ELASTOPLASTIC BEM FORMULATION USING PARTICULAR INTEGRALS

The elastoplastic, inhomogeneous differential equation given in Banerjee and Butterfield (1981) and can be written in operator notation as

$$L(\dot{u}_i) = \sigma^0_{ij,j} \tag{18}$$

where $L(\dot{u}_i)$ is a self-adjoint, homogeneous, differential operator and $\dot{\sigma}^0_{ij,j}$ is the (assumed known) inhomogeneous quantity.

The solution of the above equation can be represented as the sum of a complementary function $\dot{u}_i^c$ satisfying the homogeneous equation

$$L(\dot{u}_i^c) = 0 \tag{19}$$

and a particular integral $\dot{u}_i^p$ satisfying the inhomogeneous equation

$$L(\dot{u}_i^p) = \dot{\sigma}^0_{ij,j} \tag{20}$$

The total displacement rate $\dot{u}_i$ is expressed as $\dot{u}_i = \dot{u}_i^c + \dot{u}_i^p$.

In the theory of linear inhomogeneous differential equations, it is understood that the particular integral is not unique and any expression satisfying eqn (20) is a valid particular integral. The complementary function, when added to the particular integral, adjusts to ensure that the total solution satisfies the boundary conditions, and hence produces a unique solution to the boundary value problem.

2.4.1 Complementary Function

The (boundary only) integral equation satisfying the homogeneous part of the differential equation represents the complementary function in this procedure. The complementary function for the displacement rate at point ξ is expressed as

$$C_{ij}(\xi)\dot{u}_i^c(\xi) = \int_S [G_{ij}(x, \xi)\dot{t}_i^c(x) - F_{ij}(x, \xi)\dot{u}_i^c(x)] \, \mathrm{d}S(x) \tag{21}$$

where the $\dot{u}_i^c$ and $\dot{t}_i^c$ are the complementary functions for displacement and traction rates, respectively.

Similarly a complementary function for the interior stress rate $\dot{\sigma}_{ij}^c$ related to the homogeneous equation can be written for an interior point ξ as

$$\dot{\sigma}_{ij}^c(\xi) = \int_S [G_{kij}^\sigma(x, \xi)\dot{t}_k^c(x) - F_{kij}^\sigma(x, \xi)\dot{u}_k^c(x)] \, \mathrm{d}S(x) \tag{22}$$

The total solutions for displacement rate $\dot{u}_i$, traction rate $\dot{t}_i$, and stress rate $\dot{\sigma}_{ij}$ are

$$\dot{u}_i = \dot{u}_i^c + \dot{u}_i^p \tag{23a}$$

$$\dot{t}_i = \dot{t}_i^c + \dot{t}_i^p \tag{23b}$$

$$\dot{\sigma}_{ij} = \dot{\sigma}_{ij}^c + \dot{\sigma}_{ij}^p \tag{23c}$$

where $\dot{u}_i^p$, $\dot{t}_i^p$, and $\dot{\sigma}_{ij}^p$ are the particular integrals for displacement, traction, and stress rates, respectively.

2.4.2 Particular Integral

The particular integral is classically found via the method of undetermined coefficients, or the method of variation of parameters, or obtained by inspection of the inhomogeneous differential equation.

In the present formulation, the two- and three-dimensional particular integrals for displacement rates are related to the Galerkin vector $\dot{F}_i$ via (Fung, 1965):

$$\dot{u}_i^p = \frac{1-\nu}{\mu} \dot{F}_{i,kk} - \frac{1}{2\mu} \dot{F}_{k,ki} \tag{24}$$

where ν is Poisson's ratio.

Substituting this equation into eqn (18) renders a relationship between the Galerkin vector and the initial stress rate function:

$$\dot{F}_{i,kkjj} = \frac{1}{1-v}\dot{\sigma}^o_{ij,j} \qquad (25)$$

In subsequent steps of this derivation, it will be advantageous for the implicit expression of eqns (24) and (25) to be related by a second-order tensor, rather than a vector. Therefore, a tensor function $\dot{h}_{ij}$ is introduced such that

$$\dot{h}_{ij,mmnn} = \dot{\sigma}^o_{ij} \qquad (26)$$

Substitution of this equation into eqn (25) and simplification yields an expression for the Galerkin vector in terms of this new function:

$$\dot{F}_i = \frac{1}{1-v}\dot{h}_{ij,j} \qquad (27)$$

Finally, substituting this expression into eqn (24) yields the desired particular integral for displacement rate:

$$\dot{u}^p_i = \frac{1}{\mu}\dot{h}_{il,lkk} - \frac{1}{2\mu(1-v)}\dot{h}_{lm,ilm} \qquad (28)$$

The particular integrals for strain, stress and traction rates are found using the following relations:

$$\dot{\varepsilon}^p_{ij} = \tfrac{1}{2}(\dot{u}^p_{i,j} + \dot{u}^p_{j,i}) \qquad (29a)$$

$$\dot{\sigma}^p_{ij} = D^e_{ijkl}\dot{\varepsilon}^p_{kl} - \dot{\sigma}^o_{ij} \qquad (29b)$$

$$\dot{t}^p_i = \dot{\sigma}^p_{ij}n_j \qquad (29c)$$

It is important to note that the initial stress rate must be subtracted out of eqn (29b) in oder to produce the correct particular integral for stress rate. The complementary functions for stress and strain rates, on the other hand, are related directly by Hooke's Law.

In passing we note that the above relations for particular integrals are derived in terms of initial stress rates. An initial strain rate $\dot{\varepsilon}^o_{ij}$ formulation is possible assuming

$$\dot{h}_{ij,mmnn} = \dot{\varepsilon}^o_{ij} \qquad (30)$$

where the associated particular integral for displacement rate is

$$\dot{u}^p_i = \frac{v}{1-v}\dot{h}_{kk,ijj} + 2\dot{h}_{ij,jkk} - \frac{1}{1-v}\dot{h}_{kj,ikj} \qquad (31)$$

The particular integral equations (26)–(29) have little practical use in this implicit form. For a known initial stress or strain distribution (linear, quadratic throughout the region) eqn (31) can be used to calculate the necessary particular integral for displacements (see Henry and Banerjee, 1988a,b). For a more general distribution, however, by applying the global shape function, a concept first introduced in the context of finite elements, an explicit formulation can be developed for elastoplasticity.

2.4.3 Global Shape Function

Tensor $\dot{h}_{ij}(x)$ can be expressed in terms of a fictitious tensor density $\dot{\phi}_{ij}(\xi)$ as an infinite series using a suitable global shape function $C(x, \xi)$:

$$\dot{h}_{ml}(x) = \sum_{n=1}^{\infty} C(x, \xi_n)\dot{\phi}_{ml}(\xi_n) \tag{32}$$

where $m, l = 1, 2$ for two dimensions; $m, l = 1, 2, 3$ for three dimensions.

Several functions were considered, but the best results were obtained with the following expression:

$$C(x, \xi_n) = A_0^4(\rho^4 - b_n\rho^5) \tag{33}$$

where

A_0 is a characteristic length,

ρ is the euclidean distance between the field point x and the source point ξ_n,

b_n is a constant, chosen in a manner described by Henry (1987). For the present discussion b_n can assume the value of unity.

All distances are non-dimensionalised by a characteristic length A_0.

The unknown fictitious densities are related to the initial stress rates through eqn (26). Substituting eqn (32) into eqn (26) leads to

$$\dot{\sigma}_{lm}^{o} = \sum_{n=1}^{\infty} K(x, \xi_n)\dot{\phi}_{lm}(\xi_n) \tag{34}$$

where

$K(x, \xi_n) = C_{,mmnn}(x, \xi_n) = a - b\rho$

    $a = 8d(d + 2)$

    $b = 15b_n(d + 3)(d + 1)$

    $d = 3$ for three-dimensional analysis; $d = 2$ for two-dimensional (plane strain) analysis.

For a known initial stress rate distribution, the fictitious density functions can be determined by using this equation as described in the next section.

The particular integral for the displacement rate can be found as a function of $\phi_{lm}(\xi_n)$ by substituting eqn (32) into eqn (28):

$$\dot{u}_i^P(x) = \sum_{n=1}^{\infty} D_{iml}(x, \xi_n)\dot{\phi}_{lm}(\xi_n) \tag{35}$$

where

$$D_{iml}(x, \xi_n) = A_0\left[(c_1 + d_1\rho)(y_i\delta_{lm} + y_m\delta_{il}) + (c_2 + d_2\rho)y_l\delta_{im} + \frac{d_1}{\rho}y_iy_ly_m\right]$$

$$y_i = [x_i - (\xi_n)_i]$$

$$c_1 = \frac{-8}{2\mu(1-v)} \qquad d_1 = \frac{15b_n}{2\mu(1-v)}$$

$$c_2 = c_1 + \frac{8(d+2)}{\mu} \qquad d_2 = d_1 - \frac{15(d+3)b_n}{\mu}$$

The particular integrals for strain, stress, and traction rates are found by substituting eqn (35) into eqn (29). More specifically, the particular integral for the stress rate is expressed as

$$\dot{\sigma}_{ij}^P = \sum_{n=1}^{\infty} S_{ijlm}(x, \xi_n)\dot{\phi}_{lm}(\xi_n) \tag{36}$$

where

$$S_{ijlm}(x, \xi_n) = (e_2 + f_2\rho)\delta_{jm}\delta_{il} + (e_3 + f_3\rho)\delta_{ij}\delta_{lm} + (e_4 + f_4\rho)\delta_{im}\delta_{jl}$$

$$+ \frac{f_1}{\rho}(y_jy_m\delta_{il} + y_iy_j\delta_{lm} + y_iy_m\delta_{jl})$$

$$+ \frac{f_2}{\rho}(y_iy_l\delta_{jm} + y_jy_l\delta_{im}) + \frac{f_3}{\rho}y_ly_m\delta_{ij} + \frac{f_5}{\rho^3}y_iy_jy_ly_m$$

$$\begin{aligned} e_1 &= 2\mu c_1 & f_1 &= -f_5 = 2\mu d_1 \\ e_2 &= \mu(c_1 + c_2) & f_2 &= \mu(d_1 + d_2) \\ e_3 &= e_1 + \lambda[c_1(d+1) + c_2] & f_3 &= f_1 + \lambda[d_1(d+2) + d_2] \\ e_4 &= e_2 - a & f_4 &= f_2 + b \end{aligned}$$

The above equations are equally valid for two and three dimensions, but for two-dimensional analysis the formulation assumes the plane strain condition. The corresponding plane stress formulation can be obtained from the plane strain case by substituting the modified material constants $\bar{v}$ and $\bar{\lambda}$ into the plane strain equations in place of v and λ, respectively, where

$$\bar{v} = \frac{v}{1+v}$$

$$\bar{\lambda} = \frac{2\mu\lambda}{\lambda+2\mu}$$

2.4.4 Numerical Implementation

The boundary integral equation (21), representing the complementary function, is discretised and numerically integrated in the usual manner for a system of boundary nodes. The equations for the boundary system can be expressed for each region in matrix form as

$$\mathbf{G}\dot{\mathbf{t}}^c - \mathbf{F}\dot{\mathbf{u}}^c = 0 \qquad (37a)$$

The non-linear behaviour of the problem is included in the analysis through initial stress rates. Ultimately these initial stress rates must be determined. It is necessary to write equations for strain (or stress) rates. In matrix form, the complementary solution for the stress rate is

$$\dot{\sigma}^c = \mathbf{G}^\sigma\dot{\mathbf{t}}^c - \mathbf{F}^\sigma\dot{\mathbf{u}}^c \qquad (37b)$$

Using eqn (23), these equations can be rewritten as

$$\mathbf{G}\dot{\mathbf{t}} - \mathbf{E}\dot{\mathbf{u}} = [\mathbf{G}\dot{\mathbf{t}}^p - \mathbf{F}\dot{\mathbf{u}}^p] \qquad (38a)$$

$$\dot{\sigma} = \mathbf{G}^\sigma\dot{\mathbf{t}} - \mathbf{F}^\sigma\dot{\mathbf{u}} + [\mathbf{F}^\sigma\dot{\mathbf{u}}^p - \mathbf{G}^\sigma\dot{\mathbf{t}}^p + \dot{\sigma}^p] \qquad (38b)$$

Essentially, the particular integral solution procedure consists of evaluating the relevant particular integrals at nodal points and solving the above equation for these values and a set of appropriate boundary conditions. However, the particular integrals are a function of initial stress, and the initial stress in an inelastic analysis is unknown at the outset and must be determined as part of the solution process. Therefore, it is necessary to assemble the equation system in a manner that will admit to an inelastic solution algorithm. Both the iterative and the variable stiffness algorithms can be employed without modification if the assembly process presented below is utilised.

For the purpose of numerical evaluation, the infinite series representations of the particular integrals are truncated at a finite number of N terms:

$$\dot{u}_i^{\mathrm{p}}(x) = \sum_{n=1}^{N} D_{iml}(x, \xi_n)\dot{\phi}_{lm}(\xi_n) \tag{39a}$$

$$\dot{t}_i^{\mathrm{p}}(x) = \sum_{n=1}^{N} H_{iml}(x, \xi_n)\dot{\phi}_{lm}(\xi_n) \tag{39b}$$

$$\dot{\sigma}_{ij}^{\mathrm{p}}(x) = \sum_{n=1}^{N} S_{ijml}(x, \xi_n)\dot{\phi}_{lm}(\xi_n) \tag{39c}$$

$$\dot{\sigma}_{lm}^{o} = \sum_{n=1}^{N} K(x, \xi_n)\dot{\phi}_{lm}(\xi_n) \tag{39d}$$

The choice of N is dictated by the complexity of the problem. Generally, fictitious nodes ξ_n should be introduced at all boundary collocation nodes, and additional nodes should be added through the interior, consistent in fineness with the boundary mesh (Henry, 1987).

It should be noted that the particular integrals are derived for each region, independent of the other regions in a multi-region problem. Hence a set of (the above) equations are written for each region where plastic deformation is anticipated.

Equations for $\dot{u}_i^{\mathrm{p}}$ and $\dot{t}_i^{\mathrm{p}}$ are written at each boundary node and expressed in matrix form for each region as

$$\dot{\mathbf{u}}^{\mathrm{p}} = \mathbf{D}\dot{\boldsymbol{\phi}} \tag{40a}$$

$$\dot{\mathbf{t}}^{\mathrm{p}} = \mathbf{T}\dot{\boldsymbol{\phi}} \tag{40b}$$

in which $\mathbf{D}$ and $\mathbf{T}$ are matrices of order $(f \times M)$ by $(g \times N)$ where M is the number of boundary nodes, N is the number of terms in the series, f is the number of degrees of freedom of the analysis, and g is the number of independent stress components.

Equations for the initial stress rates are written at the N nodal points corresponding to the N ξ_n nodes via eqn (39d), and this is expressed in matrix form as

$$\dot{\boldsymbol{\sigma}}^o = \mathbf{K}\dot{\boldsymbol{\phi}} \tag{41}$$

in which $\mathbf{K}$ is a well-conditioned $(g \times N)$ by $(g \times N)$ matrix. Post-multiplying eqn (41) by $\mathbf{K}^{-1}$ yields

$$\dot{\boldsymbol{\phi}} = \mathbf{K}^{-1}\dot{\boldsymbol{\sigma}}^o \tag{42}$$

Back-substituting this equation into eqn (40) renders

$$\dot{u}^p = DK^{-1}\dot{\sigma}^o \qquad (43a)$$

$$\dot{t}^p = TK^{-1}\dot{\sigma}^o \qquad (43b)$$

Similar particular integral expressions can be written for displacement and stress at interior points of interest. The particular integral solution for stress rates, corresponding to the nodes of eqn (38b), is expressed in matrix form as

$$\dot{\sigma}^p = SK^{-1}\dot{\sigma}^o \qquad (44)$$

Substituting eqns (43) and (44) into eqn (38) and rearranging yields

$$G\dot{t} - F\dot{u} + B\dot{\sigma}^o = 0 \qquad (45a)$$

$$\dot{\sigma} = G^\sigma \dot{t} - F^\sigma \dot{u} + B^\sigma \dot{\sigma}^o \qquad (45b)$$

where

$$B = -[GT - FD]K^{-1}$$

$$B^\sigma = [G^\sigma T - F^\sigma D]K^{-1}$$

In a multi-region problem, the boundary integral equations and particular integrals are generated independently for each region, leading to a set of equations in each region similar in form to eqn (44). Interface conditions expressing the interaction of real quantities between regions are applied, and after assembling the unknown boundary quantities and corresponding coefficients on the left-hand side, and the known boundary conditions on the right, the final system can be written as

$$A^b\dot{x} = B^b\dot{y} + C^b\dot{\sigma}^o$$

$$\dot{\sigma} = A^\sigma\dot{x} + B^\sigma\dot{y} + C^\sigma\dot{\sigma}^o$$

where

$A^b, B^b, C^b, A^\sigma, B^\sigma, C^\sigma$ are block-banded matrices,
$\dot{x}$ is a vector of unknown boundary quantities,
$\dot{y}$ is a vector of known boundary conditions,
$\dot{\sigma}$ is a vector of stress rates, and
$\dot{\sigma}^o$ is a vector of initial stress rates.

The above system of equations are identical in form to those obtained by conventional volume integration, and therefore the elastoplastic solution can be obtained by employing standard elastoplastic solution procedures (iterative or variable stiffness type).

2.4.5 A Note on Efficiency

Finally, we note three time-saving features that are employed in the present implementation.

First, the complementary function for stress rates at a boundary point is represented using the boundary stress calculation, instead of the integral equation for stress rates. This circumvents the difficulty in singular integration at a boundary point.

Second, a multi-region system is used to reduce the cost of both the integration and the calculation of the particular integrals. The integration time is reduced since the surface integrals are evaluated only over the boundary of a given region, instead of over the surface of the entire body. Furthermore, particular integrals are calculated independently for each region. Thus, what would turn out to be a large matrix in a single region system is now comprised of one or more smaller matrices for which the inversions and the subsequent matrix multiplications can be performed more efficiently. Moreover, particular integrals do not have to be calculated in regions that are assumed to remain elastic.

Finally, careful observation of eqn (34) reveals that the components of the initial stress rate $\dot{\sigma}_{lm}^o$ are independently related to the components of the nodal density rate $\dot{\phi}_{lm}$. Moreover, each component has the same relation. Therefore, the tensor quantities of eqn (41) are reduced to a scalar quantity before the inversion is performed.

2.5 NUMERICAL EXAMPLES

2.5.1 Residual Stresses in a Cylindrical Rod

An axisymmetric, thermoplastic problem (Henry and Banerjee, 1987), shown in Fig. 1, regards the residual stresses in a long cylindrical rod induced by cooling. In this problem the temperature of a rod constructed of 1060 steel is raised gradually to 1250°F and then quenched in a brine spray, quickly lowering the temperature to 80°F. Experimental results obtained by Carman and Hess at Frankford Arsenal are given in Boley and Weiner (1960) for the residual stresses through the cross-section on the midplane of the rod. The Von Mises yield criterion is assumed with a yield value that varies with temperature:

$$\sigma_0(t) = \begin{cases} -14\cdot3t + 48\,710\cdot0 & \text{for } t < 400 \\ -18\cdot7t + 50\,470\cdot0 & \text{for } 400 \leqslant t \leqslant 775 \\ -46\cdot3t + 71\,900\cdot0 & \text{for } t > 775 \end{cases}$$

P.K. BANERJEE AND D.P. HENRY, JR

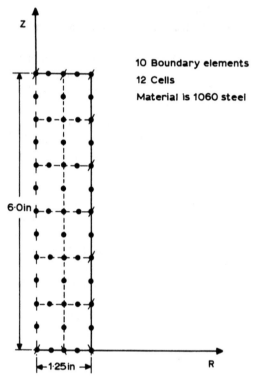

FIG. 1   Axisymmetric mesh of a cylindrical rod.

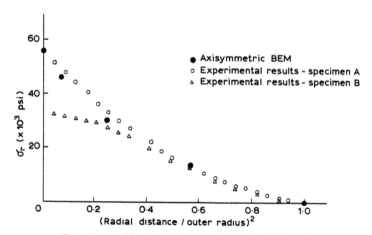

FIG. 2   Residual radial stress in a cylindrical rod.

where σ_0 is the yield stress (psi) and t is the current temperature (°F). An assumed Biot number of 10 is used in determining the rate at which heat is diffused through the rod.

Utilising the iterative plasticity algorithm, 10 quadratic boundary elements and 12 quadratic volume cells were enough to produce excellent results for this rather complex problem. Symmetry was utilised by the introduction of a roller boundary condition on the horizontal bottom face. The residual stresses along this face are given in Figs 2 and 3.

This problem was also analysed using a three-dimensional, cylindrical

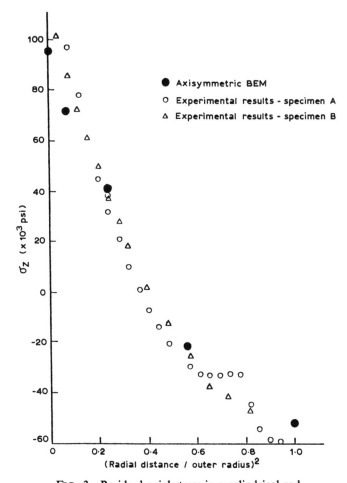

FIG. 3   Residual axial stress in a cylindrical rod.

pie-shaped model. Once again, the results closely matched the experimental solution.

2.5.2 Axisymmetric Steel Pressure Vessel

The elastoplastic analysis (Henry and Banerjee, 1987, 1988a) of a vessel subjected to internal pressure is shown in Fig. 4. The vessel, constructed of steel, has a modulus of elasticity, Poisson's ratio, and yield stress of $E = 29 \cdot 12 \times 10^6$ psi, $v = 0 \cdot 3$, and $\sigma_0 = 40\,540$ psi, respectively. The Von Mises criterion is assumed with no strain hardening. Six regions, 99 quadratic boundary elements, and 12 quadratic volume cells are used to model the body (Fig. 4). Using engineering intuition, cells must be placed in areas where yielding is anticipated. The weld connection between the spherical shell and branch is of prime concern and it is in this region that the cells are

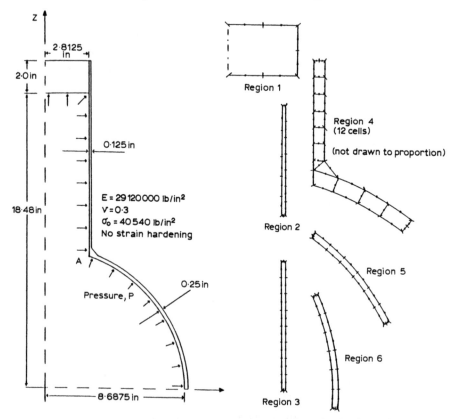

FIG. 4 Axisymmetric steel pressure vessel and discretisation by region.

situated. The other five regions are assumed to remain elastic, and at the end of the analysis it must be verified that the stresses in these regions have not violated the yield condition.

A plot of the vertical deflection of point A with increasing pressure obtained from both the variable stiffness and iterative plasticity algorithm by volume integration is shown in Fig. 5. Results obtained by Zienkiewicz (1977), using the finite element method, and test results by Dinno and Gill (1965) are also presented. The BEM results are in excellent agreement with the results obtained by FEM. The numerical solutions slightly deviate from the experimental results. This variation is due to the idealisations in the numerical analysis, such as ideal plasticity and the adoption of the Von Mises criterion. More importantly, the body is assumed homogeneous when in reality the weld and the surrounding region will exhibit greater stiffness.

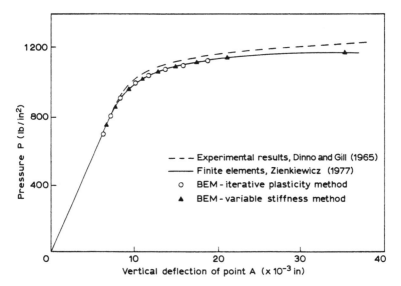

FIG. 5 Vertical deflection of point A of the pressure vessel with increasing pressure.

It should be noted that the variable stiffness method proceeds farther along the curve than the iterative procedure which does not converge at these load levels. The CPU times of the two BEM analyses were about equal, although on a virtual memory computer (HP9000) the real time of the direct plasticity method was greater than that of the iterative procedure due to excessive page faulting. The use of a multi-region system in both of the BEM analyses reduces the computational time dramatically, since each nodal equation need only be integrated over the surface in which it is contained. Furthermore, the volume integration need only be performed over the region containing cells.

In the past it has been said that BEM analysis should not be used on bodies of narrow cross-section, but through sophisticated numerical integration and the utilisation of a fine mesh, excellent results are obtained.

2.5.3 Three-dimensional Analysis of a Notch Plate

The plastic deformation of a notch plate subjected to tension is analysed under plane stress conditions. Four combinations of analyses are considered:

Particular Integral—Iterative
Particular Integral—Variable Stiffness
Volume Integral—Iterative
Volume Integral—Variable Stiffness

The material properties for the plate are:

$E = 7000 \, \text{kg/mm}^2$
$v = 0.2$
$\sigma_0 = 24.3 \, \text{kg/mm}^2$ (Von Mises yield criterion)

A 90° notch is cut out of the sides of the plate. The maximum to minimum width ratio is 2 and the thickness is 6/10 of the maximum width. A quarter of the plate is discretised in two subregions as shown in Fig. 6. The region, containing the notch, has 30 quadratic boundary elements. The inset in this figure shows six (20-node) isoparametric cells (with 68 distinct cell nodes) which are used in the volume integral-based analysis. In the particular integral analysis, particular integrals are defined using the boundary nodes and three interior nodes (corresponding to the mid-side nodes of the cells) for a total of 92 particular integral nodes. The second region has 16 quadratic boundary elements. No volume discretisation or definition of particular integrals is required in this region since it remains elastic throughout the analysis. Boundary conditions on the front and back faces are assumed traction free.

In Fig. 7, the stress–strain response on the mid-plane of the root is given for the four three-dimensional BEM analyses and compared with the two-dimensional plane stress and plane strain BEM solutions obtained by

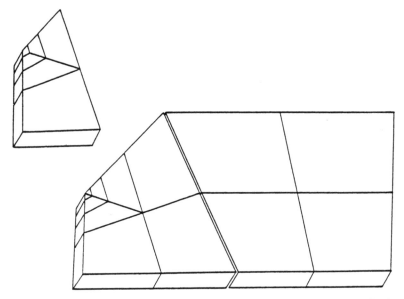

FIG. 6   Boundary and volume discretisation of a three-dimensional notch plate.

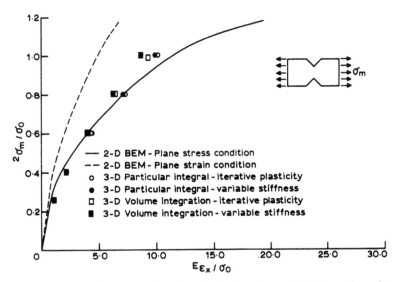

FIG. 7   Stress–strain response (on the mid-plane) at the root of a three-dimensional notch plate.

Banerjee and Raveendra (1986). The three-dimensional results are in good agreement with one another and both lie between the two two-dimensional solutions, closer to the plane stress result, as one would expect. For solutions above the load level $2\sigma_m/\sigma_0 = 1\cdot0$ requires a finer mesh around the notch, particularly in the volume cell discretisation.

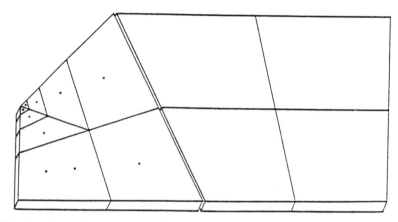

FIG. 8   Three-dimensional discretisation of a notch plate for particular integral analysis.

Figure 8 shows another notch plate mesh for a particular integral analysis. The boundary discretisation is the same, but 10 additional particular integral nodes are added in the interior (101 particular integral nodes in total). The bottom face becomes a plane of symmetry by applying a roller boundary condition. In order to keep the dimension of the plate proportional to the first analysis, the thickness of the mesh must be reduced by one-half, since symmetry is assumed by virtue of the additional roller boundary condition. The stress–strain response at the root, obtained for this mesh using the variable stiffness algorithm, is shown in Fig. 9. Results for the three nodes across the thickness of the root are compared with the two-dimensional solution (Banerjee and Raveendra, 1986). Once again, the three-dimensional results lie between the plane stress and plane strain solutions.

2.5.4 Three-dimensional Analysis of a Perforated Plate

The plastic deformation of a perforated plate in tension is analysed under plane stress conditions. The volume distribution of the initial stress is

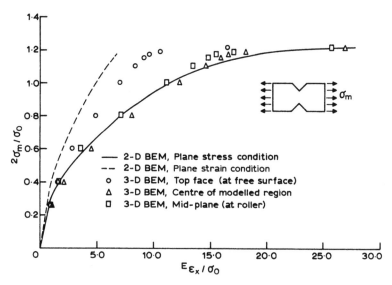

FIG. 9   Stress–strain response at the root of a notch plate, particular integral—
variable stiffness (plane stress) analysis.

represented using either 20-noded isoparametric volume cells or the
point-based particular integral. Each representation is coupled with either
the iterative or the variable stiffness solution process, leading to a total of
four distinct algorithms. The material properties for the plate are:

$$E = 7000 \, \text{kg/mm}^2$$
$$v = 0.2$$
$$\sigma_0 = 24.3 \, \text{kg/mm}^2 \text{ (Von Mises yield criterion)}$$
$$h = 224.0 \, \text{kg/mm}^2$$

The diameter of the circular hole, at the centre of the plate, is one-half the
width, and the thickness is one-fifth the width. A quarter of the plate is
discretised in two subregions, as shown in Fig. 10. The first region,
containing the root of the plate, has 30 quadratic boundary elements. For
the volume integral-based analysis the domain of this region is discretised
using nine (20-node) isoparametric cells. In the particular integral analysis,
particular integrals are defined at points corresponding to the cell nodes used
in the volume integral analysis. The second region has 23 boundary
elements. No volume discretisation or definition of particular integrals is
required in this region since it remains elastic throughout the analysis.

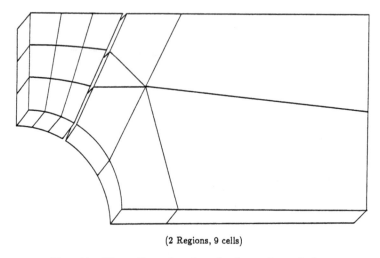

(2 Regions, 9 cells)

FIG. 10   Three-dimensional mesh of a perforated plate.

Boundary conditions on both the front and back faces are assumed traction free.

This problem was previously analysed experimentally by Theocaris and Marketos (1964) and by Zienkiewicz (1977) using the finite element method. The results obtained by the boundary element analysis are compared with

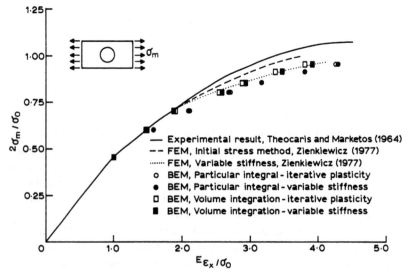

FIG. 11   Stress–strain response at the root of a three-dimensional perforated plate.

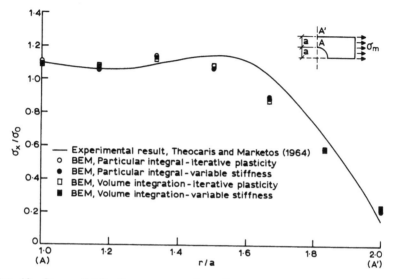

FIG. 12   Stress distribution across a three-dimensional perforated plate near collapse load (at $2\sigma_m/\sigma_0 = 0.91$).

these results in Figs 11 and 12. The stress–strain response at the root of the plate is shown in Fig. 11. The results obtained using the various BEM algorithms show good agreement with one another and with the variable stiffness FEM analysis. Differences between the iterative and variable stiffness BEM formulations are much less significant than the difference between the two FEM algorithms. The volume integral-based BEM procedure exhibits greater stiffness than the particular integral method. In Fig. 12, the stress distribution between the root and the free surface of the specimen is shown for a load of $2\sigma_m/\sigma_0 = 0.91$. Once again, excellent agreement is obtained among the four boundary element analyses. In order to evaluate the degree of convergence of the results, a mesh with 16 volume cells (or the particular integral equivalent) was studied. The results were unchanged from those shown.

The present analyses were carried out on the Cray-1 computer. The CPU times for the four algorithms were:

Particular Integral/Iterative	254 s
Particular Integral/Variable Stiffness	358 s
Volume Integral/Iterative	272 s
Volume Integral/Variable Stiffness	361 s

Of course, when the code gets optimised or vectorised, these CPU times will go down quite dramatically.

2.5.5 Two-dimensional Analysis of a Perforated Plate

The perforated plate of the previous problem is analysed using the particular integral-based, inelastic, two-dimensional, plane stress analysis. Three discretisations, shown in Fig. 13, are used to model the plate. Each mesh is divided into two subregions. The initial stress distribution in the inelastic region is defined using the particular integral representation. Twenty, 41 and 65 nodes are used, respectively, in these discretisations to define the particular integrals.

A solution is obtained using the variable stiffness method, and the results

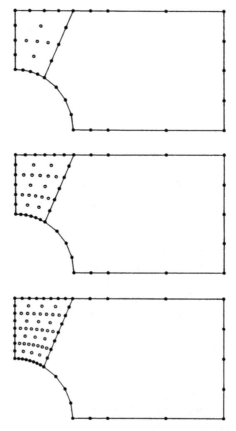

Fig. 13   Discretisation of a two-dimensional perforated plate.

are compared with the particular integral-based, variable stiffness results of the previous three-dimensional analysis. The stress–strain response at the root of the plate is shown in Fig. 14. The results obtained for the two-dimensional discretisations are in good agreement with the three-dimensional solution. The results of the more refined (65 particular integral

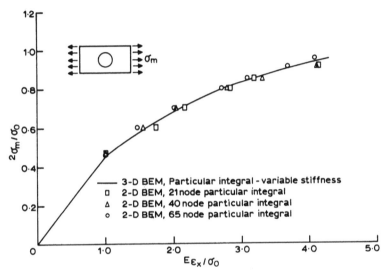

FIG. 14   Stress–strain response at the root of a perforated plate (plane stress).

nodes) mesh, however, exhibit a smoother response and follow the three-dimensional solution more closely than the other two-dimensional results. The axial stress distribution across the plate, from the root to the free edge, is shown in Fig. 15. All two-dimensional results vary from the three-dimensional solution; however, overall agreement is observed.

Finally, the computational times of the above analysis were compared using the four algorithms. The second mesh in Fig. 10 was chosen for this purpose. Nine quadratic volume cells were constructed in the notch region for the volume integral-based analysis utilising the interior points shown in the figure. The CPU times on a Hewlett-Packard 9000 (series 520) computer were:

Particular Integral/Iterative	43·5 min
Particular Integral/Variable Stiffness	49·1 min
Volume Integral/Iterative	59·1 min
Volume Integral/Variable Stiffness	72·3 min

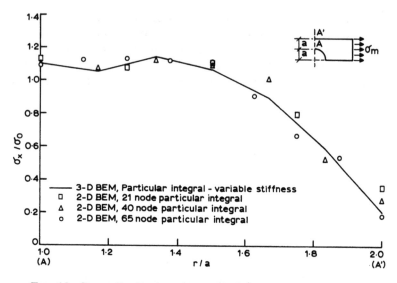

FIG. 15   Stress distribution across a perforated plate (plane stress).

2.6 CONCLUSION

A new boundary element procedure based on both volume integration and particular integrals was introduced for the elastoplastic stress analysis, and results for a range of problems were shown to correlate favourably with existing results.

The particular integrals satisfy the inhomogeneous differential equation exactly, and therefore the need for volume integrals is eliminated. The (boundary only) integral equations together with the particular integrals represent an exact statement of the problem, and any error in the BEM solution is the result of approximations and errors introduced in the numerical implementation and solution.

The computational time of a three-dimensional analysis by the particular integral method is comparable to that of volume integral formulation. For two-dimensional analysis, however, the particular integral formulation was found to be more efficient. Nevertheless, future implementation for both two- and three-dimensional analyses can be made even more computationally efficient by employing vectorising programming techniques, which are ideally suited for implementation with a large amount of matrix operations such as the particular integral-based plasticity formulation.

The present formulation introduced in this chapter can be applied to other analyses involving initial stress distribution, such as problems with initial stress induced by a 'lack of fit' stress, temperature change, inhomogeneities, geometric non-linearities, or a combination of these. Moreover, the general concept of particular integrals may be applied to BEM analyses in other areas which involve an inhomogeneous differential equation.

ACKNOWLEDGEMENTS

The work described in this chapter was made possible by the NASA contract NAS3-23697 with supplemental funding from the Structures Technology Division, Pratt & Whitney, Hartford, Connecticut and is part of BEST3D and GPBEST systems. The authors wish to express their sincere gratitude to Dr Chris Chamis, the NASA Project Manager, and Dr Edward Todd, the Pratt & Whitney Program Manager, for their support and encouragement.

REFERENCES

AHMAD, S. & BANERJEE, P.K. (1986). Free vibration analysis by BEM using particular integrals, *J. Eng. Mech. Div., Am. Soc. Civ. Eng.*, **112**(7), 682–95.

BANERJEE, P.K. & BUTTERFIELD, R. (1981). *Boundary Element Methods in Engineering Science*, McGraw-Hill, London. Also US Edition, New York, 1983; Russian Edition, Moscow, 1984.

BANERJEE, P.K. & DAVIES, T.G. (1984). Advanced implementation of boundary element methods for three-dimensional problems of elastoplasticity and viscoplasticity. Chapter 1 in: *Developments in Boundary Element Methods—3*, Eds P.K. Banerjee and S. Mukherjee, Elsevier Applied Science Publishers, London, 1–26.

BANERJEE, P.K. & HENRY, D.P., Jr (1987). Recent advances in the inelastic analysis of solids by BEM, *Proc. ASME Conf. on Advances in Inelastic Analysis*, AMD, Vol. 88, ASME, New York, 177–92.

BANERJEE, P.K. & RAVEENDRA, S.T. (1986). Advanced boundary element analysis of two- and three-dimensional problems of elastoplasticity, *Int. J. Numer. Methods Eng.*, **23**, 985–1002.

BANERJEE, P.K. & RAVEENDRA, S.T. (1987). A new boundary element formulation for two-dimensional elastoplastic analysis, *J. Eng. Mech. Div., Am. Soc. Civ. Eng.*, **113**(2), 252–65.

BANERJEE, P.K., CATHIE, D.N. & DAVIES, T.G. (1979). Two- and three-dimensional problems of elastoplasticity. Chapter 4 in: *Developments in Boundary Element Methods—1*, Eds P.K. Banerjee and R. Butterfield, Applied Science Publishers, London, 65–95.

BANERJEE, P.K., WILSON, R.B. & RAVEENDRA, S.T. (1987). Advanced applications of BEM to three-dimensional problems of monotonic and cyclic plasticity, *Int. J. Mech. Sci.*, **29**(9), 637–53.

BANERJEE, P.K., HENRY, D.P. & RAVEENDRA, S.T. (1989). Advanced inelastic analysis of solids by BEM, *Int. J. Mech. Sci.*, **31**, 309–322.

BANERJEE, P.K., AHMAD, S. & WANG, H.C. (1988a). A new BEM formulation for acoustic eigenfrequency analysis, *Int. J. Numer. Methods Eng.*, **26**, 1299–309.

BANERJEE, P.K., WILSON, R.B. & MILLER, N. (1988b). Advanced elastic and inelastic three-dimensional analysis of gas turbine engine structures by BEM, *Int. J. Numer. Methods Eng.*, **26**, 393–411.

BOLEY, B.A. & WEINER, J.H. (1960). *Theory of Thermal Stresses*, John Wiley, New York.

CATHIE, D.N. & BANERJEE, P.K. (1980). Boundary element methods for axisymmetric plasticity, in: *Innovative Numerical Methods for the Applied Engineering Science*, Eds R. Shaw, W. Pilkey, B. Pilkey, R. Wilson, A. Lakis, A. Chaudouet and C. Marino, University of Virginia Press.

CHAUDONNERET, M. (1977). Méthodes des équations intégrales appliquées à la résolution de problèmes de viscoplasticité, *J. Méchanique Appliquée*, **1**, 113–32.

DINNO, K.S. & GILL, S.S. (1965). An experimental investigation into the plastic behavior of flush nozzles in spherical pressure vessels, *Int. J. Mech. Sci.*, **7**, 817–35.

FUNG, Y.C. (1965). *Foundation of Solid Mechanics*, Prentice-Hall, Englewood Cliffs, New Jersey, 192–3.

HENRY, D.P., Jr (1987). Advanced development of the boundary element method for elastic and inelastic thermal stress analysis, Ph.D. Thesis, State University of New York at Buffalo.

HENRY, D.P., Jr & BANERJEE, P.K. (1987). A thermoplastic BEM analysis for substructured axisymmetric bodies, *J. Eng. Mech. Div., Am. Soc. Civ. Eng.*, **113**(12), 1880–900.

HENRY, D.P. & BANERJEE, P.K. (1988a). A variable stiffness type boundary element formulation for axisymmetric elastoplastic media, *Int. J. Numer. Methods Eng.*, **26**, 1005–27.

HENRY, D.P. & BANERJEE, P.K. (1988b). A new BEM formulation for thermoelasticity by particular integrals, *Int. J. Numer. Methods Eng.*, **26**(9), 2061–78.

HENRY, D.P. & BANERJEE, P.K. (1988c). A new BEM formulation for elastoplasticity by particular integrals, *Int. J. Numer. Methods Eng.*, **26**(9), 2079–96.

HENRY, D.P., Jr, PAPE, D.A. & BANERJEE, P.K. (1987). A new axisymmetric BEM formulation for body forces using particular integrals, *J. Eng. Mech. Div., Am. Soc. Civ. Eng.*, **113**(5), 671–88.

KUMAR, V. & MUKHERJEE, S. (1975). A boundary integral equation formulation for time dependent inelastic deformation in metals, *Int. J. Mech. Sci.*, **19**(12), 713–24.

PAPE, D.A. & BANERJEE, P.K. (1987). Treatment of body forces in 2-D elastostatic BEM using particular integrals, *J. Appl. Mech., ASME*, **54**, 866–71.

RAVEENDRA, S.T. (1984). Advanced development of BEM for two- and three-dimensional nonlinear analysis, Ph.D. Dissertation, State University of New York at Buffalo.

RICCARDELLA, P. (1973). An implementation of boundary-integral technique for planar problems of elasticity and elastoplasticity, Ph.D. Thesis, Carnegie–Mellon University, Pittsburgh, Pennsylvania.

SWEDLOW, J.L. & CRUSE, T.A. (1971). Formulation of boundary integral equations for three-dimensional elastoplastic flow, *Int. J. Solids Struct.*, **7**, 1673–83.

TELLES, J.C.F. (1983). The boundary element method applied to inelastic problems, Lecture notes in engineering, Springer–Verlag, New York.

THEOCARIS, P.S. & MARKETOS, E. (1964). Elasto-plastic analysis of perforated thin strips of a strain hardening material, *J. Mech. Phys. Solids*, **12**, 377–90.

WANG, H.C. & BANERJEE, P.K. (1988). Axisymmetric free-vibration problems by the boundary element method, *J. Appl. Mech., ASME*, **55**, 437–42.

WATSON, J.O. (1979). Advanced implementation of the boundary element method for two- and three-dimensional elastostatics. Chapter 3 in: *Developments in Boundary Element Methods—1*, Eds P.K. Banerjee and R. Butterfield, Applied Science Publishers, London, 31–63.

ZIENKIEWICZ, O.C. (1977). *The Finite Element Method*, McGraw-Hill, London.

Chapter 3

ADVANCED DEVELOPMENT OF BEM FOR ELASTIC AND INELASTIC DYNAMIC ANALYSIS OF SOLIDS

P.K. BANERJEE, S. AHMAD and H.C. WANG

*Department of Civil Engineering, State University of New York at Buffalo,
USA*

SUMMARY

*Direct Boundary Element formulations and their numerical implementation
for periodic and transient elastic as well as inelastic transient dynamic
analyses of two-dimensional, axisymmetric and three-dimensional solids are
presented. The inelastic formulation is based on an initial stress approach and
is the first of its kind in the field of Boundary Element Methods. This
formulation employs the Navier–Cauchy equation of motion, Graffi's dynamic
reciprocal theorem, Stokes' fundamental solution, and the Divergence theorem,
together with kinematical and constitutive equations to obtain the pertinent
integral equations of the problem in the time domain within the context of the
small displacement theory of elastoplasticity.*

*The dynamic (periodic, transient as well as non-linear transient) formul-
ations have been applied to a range of problems. The numerical formulations
presented here are included in the BEST3D and GPBEST systems.*

3.1 INTRODUCTION

For linear elastic problems under static or dynamic loading, the BEM has
a number of distinct advantages over the FEM, as explained in Banerjee
and Butterfield (1981), Banerjee *et al.* (1986), Banerjee and Ahmad (1985)
and Ahmad (1986). Some of these advantages for linear dynamic problems

77

are: (i) accurate and efficient solutions of problems involving semi-infinite or infinite domains, because the radiation condition at infinity is automatically satisfied by the Stokes' fundamental solution: (ii) reduction of the dimensionality of the problem by one, resulting in a discretisation of just the surface of the domain instead of the whole domain (as required in the FEM).

However, for problems with material non-linearity, in addition to the boundary discretisation, an interior discretisation of a small part of the domain where non-linear behaviour is expected becomes necessary. Since the number of unknowns in the resulting algebraic system of equations depends only on the boundary discretisation, a considerable reduction in the size of the problem is still achieved. The accuracy of stress computation is of great importance in a non-linear analysis, because the inelastic strain rates to be calculated during the analysis are proportional to stress ratios. This advantage of the BEM is evident in elastoplasticity and viscoplasticity under static or quasistatic loadings, as explained in Banerjee and Butterfield (1981), Mukherjee (1982), etc.

All previous work on dynamic anaysis by the BEM is based on the assumption of linear elastic behaviour, and most assumes steady-state (time-harmonic) conditions. For a truly transient process, it is thus mandatory to consider time-dependent response and non-linear behaviour.

The inelastic analysis described in this chapter, previously reported in a non-archival report for NASA contract NAS3-23697 by Wilson and Banerjee (1986), in the PhD dissertation of Ahmad (1986) and through private communication to Manolis and Beskos (1986), represents the first effort towards the development of a general numerical methodology for solving non-linear transient dynamic problems by using the BEM. This work essentially combines the knowledge and experience from our recent work in time-domain transient elastodynamics (Banerjee and Ahmad, 1985; Banerjee et al., 1986; Ahmad and Banerjee, 1988b) and static elastoplasticity (Banerjee and Davies, 1984; Banerjee and Raveendra, 1986; Henry, 1987). In addition major applications of linear (periodic and transient) dynamic analysis are also described.

3.2 BEM FOR PERIODIC DYNAMIC ANALYSIS

By using the dynamic reciprocal theorem, the boundary integral representation for periodic dynamics can be written in the frequency domain as

(Banerjee and Butterfield, 1981):

$$c_{ij}(\xi)u_j(\xi, \omega) = \int_S [G_{ij}(x, \xi, \omega)t_i(x, \omega) - F_{ij}(x, \xi, \omega)u_i(x, \omega)] \, \mathrm{d}S(x) \quad (1)$$

where S is the surface enveloping the body and the fundamental solutions for G_{ij} and F_{ij} are defined in Banerjee and Ahmad (1985), Ahmad (1986), Banerjee et al. (1988) and Ahmad and Banerjee (1988a). It should be noted here that although the functions G_{ij} and F_{ij} become identical to their static counterparts as ω tends to zero, it is important to evaluate this limit carefully because of the presence of ω in the denominator.

Once the boundary solution is obtained, an interior version of eqn (1) can also be used to find the interior displacements; and the interior stresses can then be obtained using the strain–displacement and stress–strain relations as (Banerjee and Ahmad, 1985):

$$\sigma_{jk}(\xi, \omega) = \int_S [G_{ijk}^\sigma(x, \xi, \omega)t_i(x, \omega) - F_{ijk}^\sigma(x, \xi, \omega)u_i(x, \omega)] \, \mathrm{d}S(x) \quad (2)$$

The functions G_{ijk}^σ and F_{ijk}^σ of the above equation are listed in Banerjee and Ahmad (1985), Ahmad (1986), Banerjee et al. (1988) and Ahmad and Banerjee (1988a).

The stresses at the surface can be calculated by combining the constitutive equations, the directional derivatives of the displacement vector and the values of field variables in an accurate matrix formulation described later.

The boundary integral formulations (1) and (2) can also take account of internal viscous dissipation of energy (damping) using complex elastic moduli in the usual manner.

3.2.1 Numerical Implementation

The boundary integral statements (1) and (2) of the governing differential equation of the problem represent an exact statement of the problem. Unlike the finite element method where there is an energy minimisation implied in a discretised system, the boundary element discretised system does not use any energy principle; only the exact representations (1) and (2) make such a statement. Such a system must therefore be implemented as accurately as possible, otherwise the benefits of the formulation cannot be realised. Essential features of the current implementation (which are similar to those outlined in Chapter 1 of this book) are briefly described below:

(1) The geometry and the boundary values of displacements and tractions are represented by isoparametric quadratic shape functions.

(2) The numerical integration is carried out to a preselected 3–5 digit precision uniformly for all integrals (both singular and non-singular integrals). This requires extensive self-adaptive subdivision of elements to reflect the behaviour of the kernel-shape-function–Jacobian products. These are discussed extensively in the boundary element literature.

(3) Substructuring facilities have been provided, since substructuring allows the modelling of layering of soils and soil-structure interaction problems. Essentially substructuring technique allows a problem geometry to be modelled as an assembly of several generic modelling regions (GMR). The GMRs are assembled together by enforcing the continuity of displacements and equilibrium of tractions across common boundary elements.

(4) After the system equations are obtained in the form

$$[A]\{x\} = [B]\{y\} \tag{3}$$

an out-of-core complex solver developed using software from LINPAC is used to obtain the unknown boundary values. In this solver, in order to minimise the time requirements, the solution process is carried out using the block form of the matrix.

(5) The numerical solution of the transient dynamic problem can also be carried out in Laplace transform domain. This consists of a series of solutions to a steady-state dynamic problem for a number of discrete values of the transformed parameter s. The final solution is then obtained by a numerical inversion of the transformed domain solutions to the time domain. Thus, the steady-state dynamic boundary element formulation and its numerical implementation presented in previous sections are also valid for Laplace-transformed transient dynamic analysis (Ahmad and Manolis, 1987; Ahmad and Banerjee, 1988a).

3.2.2 General Axisymmetric Formulation

The term 'general axisymmetric' here means that the geometry is axisymmetric but the loading conditions may be non-axisymmetric. Essentially the method used involves the well-known Fourier series expansion of boundary tractions and displacements to decompose the problem into a series of uncoupled subproblems, each of which denotes the distribution of displacement and traction fields in the corresponding circumferential order.

The boundary integral equation for three-dimensional periodic dynamics is given as

$$c_{ij}(\xi)u_i(\xi, \omega) = \int_S [G_{ij}(x, \xi, \omega)t_i(x, \omega) - F_{ij}(x, \xi, \omega)u_i(x, \omega)] \, dS(x) \quad (4)$$

Considering the symmetric case,

$$u_r = \sum {}^c u_r^n \cos n\theta \qquad t_r = \sum {}^c t_r^n \cos n\theta$$
$$u_\theta = \sum {}^s u_\theta^n \sin n\theta \qquad t_\theta = \sum {}^s t_\theta^n \sin n\theta$$
$$u_z = \sum {}^c u_z^n \cos n\theta \qquad t_z = \sum {}^c t_z^n \cos n\theta$$

then the boundary equation for general axisymmetric periodic elasto-dynamics can be derived as (Rizzo and Shippy, 1979):

$$c_{ij}(\xi)u_i^n(\xi, \omega) = \int_L [G_{ij}^n(x, \xi, \omega)t_i^n(x, \omega) - F_{ij}^n(x, \xi, \omega)u_i^n(x, \omega)] \, dS(x) \quad (5)$$

where in matrix notation

$$\mathbf{u_i^n} = \begin{Bmatrix} {}^c u_r^n \\ {}^s u_\theta^n \\ {}^c u_z^n \end{Bmatrix} \qquad \mathbf{t_i^n} = \begin{Bmatrix} {}^c t_r^n \\ {}^s t_\theta^n \\ {}^c t_z^n \end{Bmatrix}$$

$$\mathbf{G_{ij}^n} = \begin{bmatrix} G_{rr}^{nc} & G_{r\theta}^{ns} & G_{rz}^{nc} \\ -G_{\theta r}^{ns} & G_{\theta\theta}^{nc} & -G_{\theta z}^{ns} \\ G_{zr}^{nc} & G_{z\theta}^{ns} & G_{zz}^{nc} \end{bmatrix} \qquad \mathbf{F_{ij}^n} = \begin{bmatrix} F_{rr}^{nc} & F_{r\theta}^{ns} & F_{rz}^{nc} \\ -F_{\theta r}^{ns} & F_{\theta\theta}^{nc} & -F_{\theta z}^{ns} \\ F_{zr}^{nc} & F_{z\theta}^{ns} & F_{zz}^{nc} \end{bmatrix}$$

and

$$G_{ij}^{nc} = \int_0^{2\pi} G_{ij}(x, \xi, \omega) \cos n\theta \, d\theta \qquad F_{ij}^{nc} = \int_0^{2\pi} F_{ij}(x, \xi, \omega) \cos n\theta \, d\theta$$

$$G_{ij}^{ns} = \int_0^{2\pi} G_{ij}(x, \xi, \omega) \sin n\theta \, d\theta \qquad F_{ij}^{ns} = \int_0^{2\pi} F_{ij}(x, \xi, \omega) \sin n\theta \, d\theta$$

$$\theta = \theta_x - \theta_\xi$$

Discretising the generator of the boundary surface, L, and letting ζ approach the boundary surface, after the necessary circumferential integration, we are able to form

$$[G]^n\{t\}^n - [F]^n\{u\}^n = \{0\} \tag{6}$$

After the boundary conditions are expressed in Fourier series and applied, the coefficients in the Fourier series expansion of displacement and traction fields can be solved for every circumferential order separately. Finally, the real displacement and traction fields are found by superposing the result in each circumferential order according to the Fourier series.

If asymmetric terms are considered, the following modifications have to be made:

$$u_r = \sum {}^s u_r^n \sin n\theta \qquad t_r = \sum {}^s t_r^n \sin n\theta$$

$$u_\theta = \sum {}^c u_\theta^n \cos n\theta \qquad t_\theta = \sum {}^c t_\theta^n \cos n\theta$$

$$u_z = \sum {}^s u_z^n \sin n\theta \qquad t_z = \sum {}^s t_z^n \sin n\theta$$

$$\mathbf{u}_i^n = \begin{Bmatrix} {}^s u_r^n \\ {}^c u_\theta^n \\ {}^s u_z^n \end{Bmatrix} \qquad \mathbf{t}_i^n = \begin{Bmatrix} {}^c t_r^n \\ {}^s t_\theta^n \\ {}^c t_z^n \end{Bmatrix}$$

$$\mathbf{G}_{ij}^n = \begin{bmatrix} G_{rr}^{nc} & -G_{r\theta}^{ns} & G_{rz}^{nc} \\ G_{\theta r}^{ns} & G_{\theta\theta}^{nc} & G_{\theta z}^{ns} \\ G_{zr}^{nc} & -G_{z\theta}^{ns} & G_{zz}^{nc} \end{bmatrix} \qquad \mathbf{F}_{ij}^n = \begin{bmatrix} F_{rr}^{nc} & -F_{r\theta}^{ns} & F_{rz}^{nc} \\ F_{\theta r}^{ns} & F_{\theta\theta}^{nc} & F_{\theta z}^{ns} \\ F_{zr}^{nc} & -F_{z\theta}^{ns} & F_{zz}^{nc} \end{bmatrix}$$

In the numerical implementation the kernel functions involved were decomposed into a static part and a dynamic part. The circumferential integration of the static part was accomplished analytically leading to elliptic integrals. For the dynamic part, however, the integration was carried out numerically. Up to four circumferential orders were implemented.

3.3 TRANSIENT BOUNDARY INTEGRAL FORMULATION

The direct boundary integral formulation for a general, transient, elasto-dynamic problem can be constructed by combining the fundamental point-force solution (Stokes' solution) of the governing equations of motion (Navier–Cauchy equations) with Graffi's (1947) dynamic reciprocal theorem.

Details of this construction can be found in Banerjee and Butterfield (1981). For zero initial conditions and zero body forces, the boundary integral formulation for transient elastodynamics reduces to:

$$c_{ij}(\xi)u_i(\xi, T) = \int_S [G_{ij}(x, \xi, T)^* t_i(x, T)$$

$$- F_{ij}(x, \xi, T)^* u_i(x, T)]\, dS(x) \qquad (7)$$

where

$$G_{ij}^* t_i = \int_0^T G_{ij}(x, T; \xi, \tau) t_i(x, \tau)\, d\tau$$

$$\qquad (8)$$

$$F_{ij}^* u_i = \int_0^T F_{ij}(x, T; \xi, \tau) u_i(x, \tau)\, d\tau$$

are Riemann convolution integrals and ξ and x are the positions of the receiver (field point) and the source (source point). The fundamental solutions G_{ij} and F_{ij} are the displacements u_i and tractions t_i at a point x at time T due to a unit force vector applied at a point ξ at a preceding time τ and are defined in Ahmad (1986) and Ahmad and Banerjee (1988b).

Equation (7) represents an exact formulation involving integration over the surface as well as the time history. It should also be noted that this is an implicit time-domain formulation because the response of time T is calculated by taking into account the history of surface tractions and displacements up to and including the time T. Furthermore, eqn (7) is valid for both regular and unbounded domains.

Once the boundary solution is obtained, the stresses at the boundary can be calculated by combining the constitutive equations, the directional derivatives of the displacement vector and the values of field variables in an accurate matrix formulation. For calculating displacements at interior points, eqn (7) can be used with $c_{ij} = \delta_{ij}$ and the interior stresses can be obtained from

$$\sigma_{jk}(\xi, T) = \int_S [G_{ijk}^\sigma(x, \xi, T)^* t_i(x, T) - F_{ijk}^\sigma(x, \xi, T)^* u_i(x, T)]\, dS(x) \quad (9)$$

where the asterisk (*) again indicates Riemann convolution.

The functions G^σ_{ijk} and F^σ_{ijk} of the above equation are defined in Ahmad and Banerjee (1988b). These functions contain first- and second-order derivatives with respect to time. Therefore, the use of constant temporal variation of the field variables, a feature of all previous work, produces unsatisfactory results for interior stresses.

3.3.1 Numerical Implementation

Since the present implementation is a part of BEST3D (and also of GPBEST), details of the numerical schemes for surface discretisation, spatial integration, assembly and solution of system equations can be found in Chapter 1 of this volume. Only a brief description of temporal integration and the time-marching scheme used in the present work are presented below. Further details can be seen in Banerjee and Ahmad (1985), Banerjee et al. (1986) and Ahmad and Banerjee (1988b).

In order to obtain the transient response at a time T_N, the time axis is discretised into N equal time intervals, i.e.

$$T_N = \sum_{n=1}^{N} n\Delta T \qquad (10)$$

where ΔT is the time step.

Utilising eqn (10), eqn (7) can be written as:

$$c_{ij}u_i(\xi, T_N) - \int_{T_{N-1}}^{T_N} \int_S [G_{ij}t_i - F_{ij}u_i]\, dS\, d\tau$$

$$= \int_{\tau=0}^{T_{N-1}} \int_S [G_{ij}t_i - F_{ij}u_i]\, dS\, d\tau \qquad (11)$$

where the integral on the right-hand side is the contribution due to past dynamic history.

It is of interest that eqn (11), like eqn (7), still remains an exact formulation of the problem since no approximation has yet been introduced. However, in order to solve eqn (11), one has to approximate the time variation of the field quantities in addition to making the usual approximation of spatial variation. For this purpose, a linear interpolation function is used which is described with the resulting time-stepping algorithms as follows.

The field variables (i.e. displacements, tractions or stresses) are assumed to vary linearly during a time step, i.e.

$$u_i(x, \tau) = \sum_{n=1}^{N} [M_I^n u_i^{n-1}(x) + M_F^n u_i^n(x)] \tag{12}$$

$$t_i(x, \tau) = \sum_{n=1}^{N} [M_I^n t_i^{n-1}(x) + M_F^n t_i^n(x)] \tag{13}$$

where N is the total number of time steps; $u_i^n(x)$ and $t_i^n(x)$ represent the spatial variation of u_i and t_i, respectively, at time T_n; and M_I and M_F are the temporal interpolation functions related to local time nodes I and F, and are of the form:

$$M_I^n = \frac{T_n - \tau}{\Delta T} \phi_n(\tau) \qquad M_F^n = \frac{\tau - T_{n-1}}{\Delta T} \phi_n(\tau) \tag{14}$$

where

$\phi_n(\tau) = 1$ for $(n-1)\Delta T \leqslant \tau \leqslant n\Delta T$, and

$\quad = 0$ otherwise; and

$\phi_n(\tau) = [H\{\tau - (n-1)\Delta T\} - H\{\tau - n\Delta T\}]$

H being the Heaviside function.

For illustration purposes, consider the boundary integral equation (11) for the first time step, i.e.

$$c_{ij} u_i(\xi, T_1) - \int_{T_0}^{T_1} \int_S [G_{ij} t_i - F_{ij} u_i] \, dS \, d\tau = 0 \tag{15}$$

The time integration in eqn (15) after using eqn (14) is done analytically and the spatial integration is performed numerically (Banerjee et al., 1986). This analytical temporal integration is described in Ahmad and Banerjee (1988b).

After the integrations and the usual assembly process, the resulting system equation has the form

$$[A_F^1]\{X^1\} - [B_F^1]\{Y^1\} + [A_I^1]\{X^0\} - [B_I^1]\{Y^0\} = \{0\} \tag{16}$$

where:

[A]   and [B] are the matrices related to the unknown and known field
      quantities, respectively;

{X}   and {Y} are the vectors of unknown and known field quantities,
      respectively;

for $\{X\}$ and $\{Y\}$ the superscript refers to the time step;

for $[A]$ and $[B]$ the superscript denotes the time step at which they are calculated, and the subscript denotes the local time nodes (I or F) during that time-stepping interval.

Since all the unknowns at time $T=0$ are assumed to be zero, eqn (16) reduces to

$$[A_F^1]\{X^1\}=[B_F^1]\{Y^1\}+[B_I^1]\{Y^0\} \tag{17}$$

Now consider the boundary integral equation for the second time step, i.e.

$$c_{ij}u_i(\xi, T_2)-\int_{T_1}^{T_2}\int_S [G_{ij}t_i-F_{ij}u_i]\,dS\,d\tau$$

$$=\int_{T_0}^{T_1}\int_S [G_{ij}t_i-F_{ij}u_i]\,dS\,d\tau \tag{18}$$

If the time interval (T_2-T_1) is the same as (T_1-T_0) the resulting kernels in the left-hand sides of eqns (15) and (18) become identical. This is due to the time translation properties of the fundamental solutions G_{ij} and F_{ij}, which contain time functions with arguments $(T-\tau)$, and therefore the convoluted integral corresponding to the interval $T_1\leqslant\tau\leqslant T_2$ with $T=T_2$ is identical to that of the interval $T_0\leqslant\tau\leqslant T_1$ with $T=T_1$.

The right-hand side of eqn (18) is evaluated at time $T=T_2$ with the time integration over the interval T_0 to T_1 and thus provides the effects of the dynamic history of the first time interval on the current time node (i.e. T_2).

Now, the resulting system equation for this time node (T_2) is of the form

$$[A_F^1]\{X^2\}-[B_F^1]\{Y^2\}+[A_I^1]\{X^1\}-[B_I^1]\{Y^1\}$$
$$=-\{[A_F^2]\{X^1\}-[B_F^2]\{Y^1\}+[A_I^2]\{X^0\}-[B_I^2]\{Y^0\}\} \tag{19}$$

Equation (19) can be rearranged such that

$$[A_F^1]\{X^2\}=[B_F^1]\{Y^2\}-[A_I^1+A_F^2]\{X^1\}$$
$$+[B_I^1+B_F^2]\{Y^1\}+[B_I^2]\{Y^0\} \tag{20}$$

In the above equation, all the quantities on the right-hand side are known. Therefore the unknown vector $\{X^2\}$ at time T_2 can be obtained by solving the above equation.

Thus, for the Nth time step, the boundary integral equation (11) can be written in a discretised form as

$$[A_F^1]\{X^N\} - [B_F^1]\{Y^N\} = - \sum_{n=2}^{N} [[A_F^n + A_I^{n-1}]\{X^{N-n+1}\}$$
$$- [B_F^n + B_I^{n-1}]\{Y^{N-n+1}\}] + [B_I^n]\{Y^0\} \quad (21)$$

or

$$[A_F^1]\{X^N\} = [B_F^1]\{Y^N\} + \{R^N\} \quad (22)$$

where vector $\{R^N\}$ is the effect of the past dynamic history on the current time node.

The above equation can be solved to find the unknown vector $\{X^N\}$ at time T_N. It may appear at first glance that a prodigious calculation of coefficients is involved. However, a closer examination will reveal that:

(1) If the time step size is constant, the $[A_F^1]$ and $[B_F^1]$ matrices do not change from time step to time step.
(2) For each time step, a new $\{R^N\}$ vector needs to be formed. This involves the evaluation of a new set of coefficient matrices $[A_I^n], [B_I^n], [A_F^n]$ and $[B_F^n]$ involving the effects of the dynamic history of the first time interval on the current time node. Eventually, however, this contribution to $\{R^N\}$ reduces to zero and from that point onwards no new coefficients need to be evaluated.

Finally, it is of interest to note that if the linear temporal shape functions M_I^n and M_F^n are replaced by $M_I^n = M_F^n = 0.5\phi_n(\tau)$, the present time stepping scheme reduces to a constant temporal variation scheme. In this constant temporal variation scheme, the averaged value of a field variable between the two local time nodes (I and F) is taken as the representative value for the field variable during a time step. This averaging yields more accurate results (Banerjee *et al.*, 1986) than that obtained without averaging, although the latter approach has been invariably used by most researchers in the past.

3.4 BEM FORMULATIONS FOR DYNAMIC PLASTICITY

Governing differential equations for the dynamic plasticity can be written as (Ahmad and Banerjee, 1988c)

$$(\lambda + \mu)\Delta u_{j,ij} + \mu\Delta u_{i,jj} + \Delta f_i = \Delta\sigma_{ij,j}^0 + \rho\Delta u_i \quad (23)$$

The direct boundary integral formulation for an inelastic transient dynamic problem, based on an initial stress approach, can be constructed by following a procedure similar to the one that has been used for an inelastic static problem (Banerjee and Butterfield, 1981). Under zero initial conditions and zero body forces, the boundary integral equation for inelastic transient dynamics can be expressed as

$$c_{ij}(\xi)\Delta u_i(\xi, T) = \int_S [G_{ij}(x, \xi, T)*\Delta t_i(x, T) - F_{ij}(x, \xi, T)*\Delta u_i(x, T)] \, dS(x)$$

$$+ \int_V B_{iij}(x, \xi, T)*\Delta\sigma_{ii}^0(x, T) \, dV(x) \tag{24}$$

where an asterisk (*) denotes Riemann convolution,

$$G_{ij}*\Delta t_i = \int_0^T G_{ij}(x, T; \xi, \tau)\Delta t_i(x, \tau) \, d\tau$$

and ξ and x are the space positions of the receiver (field point) and the source (source point), respectively; and where $B_{iij} = G_{ij,l}$.

Since constitutive equations expressing the material behaviour in the inelastic range are only valid in incremental form, it is necessary for the incremental boundary integral equations for the inelastic transient dynamic problem to be integrated over the entire loading history. In the elastodynamic problem, this is automatically achieved through the Riemann convolution of incremental boundary displacement, and surface traction, as well as the initial stress histories as exemplified in eqn (24).

3.4.1 Stresses at Interior Points

The integral equations for stress increment at an interior point can be obtained by taking derivatives of the boundary integral equation for an interior point (i.e. eqn (24) with $c_{ij} = \delta_{ij}$), and by using incremental constitutive and kinematic equations (Banerjee and Raveendra, 1986). This equation has the form:

$$\Delta\sigma_{jk}(\xi, T) = \int_S [G_{ijk}^\sigma(x, \xi, T)*\Delta t_i(x, T) - F_{ijk}^\sigma(x, \xi, T)*\Delta u_i(x, T)] \, dS$$

$$+ \int_V B_{iijk}^\sigma(x, \xi, T)*\Delta\sigma_{ii}^0(x, T) \, dV + J_{iijk}\Delta\sigma_{ii}^0(\xi, T) \tag{25}$$

The functions G^σ_{ijk}, F^σ_{ijk}, B^σ_{iljk} and J_{iljk} are defined in Ahmad (1986) and also in Ahmad and Banerjee (1988c).

In eqn (25), the volume integral is strongly singular, therefore it must be evaluated in the sense of $(V - V_\varepsilon)$ with limit $V_\varepsilon \to 0$. Because of the singularity, the tensor J_{iljk} is the jump or discontinuity tensor obtained from the analytical treatment of the integral over V_ε. This jump term is the same as that of static plasticity and is independent of the size of the exclusion V_ε, provided the initial stress distribution is homogeneous within V_ε (see Banerjee and Davies (1984), Banerjee and Raveendra (1986), and Henry (1987)).

3.4.2 Stresses at Boundary Points

The stresses (incremental) at the boundary points and at points very close to the boundary cannot be evaluated by using eqn (25), due to the strongly singular nature of the integrals involved. However, these can be obtained by combining the constitutive equation, the directional derivatives of the displacement vector and the values of the field variables in an accurate matrix formulation. Using this scheme, the incremental stresses at boundary point ξ^b can be obtained by coupling the following set of equations:

$$\Delta\sigma_{ij}(\xi^b, T) = [\lambda\delta_{ij}\Delta u_{m,m}(\xi^b, T) + \mu\{\Delta u_{i,j}(\xi^b, T) + \Delta u_{j,i}(\xi^b, T)\}]$$
$$- \Delta\sigma^0_{ij}(\xi^b, T) \qquad (26a)$$

$$\Delta t_i(\xi^b, T) = \Delta\sigma_{ij}(\xi^b, T)n_j(\xi^b) \qquad (26b)$$

$$\frac{\partial\Delta u_i}{\partial\eta_\alpha} = \frac{\partial\Delta u_i}{\partial\xi^b_j}\frac{\partial\xi^b_j}{\partial\eta_\alpha} \qquad (26c)$$

where η_α is a set of local axes at the field point (ξ^b).

3.4.3 Constitutive Model

In dynamic plasticity, the choice of an appropriate constitutive model depends largely on the material properties and the loading conditions of the problem in hand. For this reason many constitutive models have been proposed for dynamic plasticity. However, for simplicity, the Von Mises model with isotropic variable hardening is used here (Hill, 1950).

In this model, the behaviour in the elastoplastic region is governed by the stress–strain relations:

$$\Delta\sigma_{ij} = 2\mu\left[\Delta\varepsilon_{ij} + \frac{v}{1-v}\delta_{ij}\Delta\varepsilon_{kk} - \frac{3S_{ij}S_{kl}}{2\sigma_0^2(1+H/3\mu)}\Delta\varepsilon_{kl}\right] \qquad (27)$$

where

$\Delta\sigma_{ij} = D^{ep}_{ijkl}\Delta\varepsilon_{kl}$ = incremental stress tensor,

D^{ep}_{ijkl} = incremental elastoplastic material modulus,

μ = elastic shear modulus,

v = Poisson's ratio,

σ_0 = uniaxial yield stress = $\sqrt{3S_{ij}S_{ij}/2}$,

S_{ij} = deviatoric stress = $\sigma_{ij} - \delta_{ij}\sigma_{kk}/3$, and

H = plastic-hardening modulus, the current slope of the uniaxial plastic stress–strain curve.

The present implementation is such that any other constitutive model can be included without major difficulty.

3.4.4 Numerical Implementation

The boundary integral equations (24) and (25) provide the formal basis for developing the algorithm for the inelastic transient dynamic analysis. However, the initial stresses $\Delta\sigma^0_{ij}$ defined in these equations are not known *a priori* and have to be evaluated by satisfying the constitutive relations (27).

Since the volume integrals of eqns (24) and (25) related to inelastic stress vanish except in regions of non-linear material response, approximations of geometry and field quantities are required only where non-linearity is expected. In the present work, isoparametric 20-noded volume cells are used for approximating the geometry and the variation of initial stresses such that

$$x_i = N_\beta(\eta)(\bar{X}_i)_\beta \qquad (28a)$$

$$\sigma^0_{ij} = N_\beta(\eta)(\bar{\sigma}^0_{ij})_\beta \qquad (28b)$$

where

x_i are cartesian coordinates,

$(\bar{X}_i)_\beta$ are nodal coordinates of the volume cell,

N_β is a quadratic shape function for the volume cell,

β represents the nodal points of the volume cell, and

$^-$ (bar) denotes nodal quantities.

The volume integral of eqn (24) can then be represented as

$$\int_0^T \int_V B_{ilk}(x, T; \xi^b, \tau) \Delta \sigma_{il}^0(x, \tau) \, dV \, d\tau$$

$$= \sum_{m=1}^{L} \int_0^T \int_{V_m} B_{ilj}[x^m(\eta), T; \xi^b, \tau] N_\beta(\eta) \, dV_m (\Delta \bar{\sigma}_{il}^0)_\beta^m \, d\tau \qquad (29)$$

where:

ξ^b is the field point on the boundary (boundary node),

$x^m(\eta)$ is the point in cell m,

$(\Delta \bar{\sigma}_{il}^0)_\beta^m$ are the nodal values of incremental initial stress of the mth cell,

V_m is the volume of the mth cell, and

L is the total number of cells.

The volume integral of the interior stress equation (25) can also be expressed in a similar manner. The resulting integrations are carried out in much the same manner as in the static problem (Banerjee and Raveendra, 1986) except that more volume subsegmentation is used for the present case.

3.4.5 Time-marching Scheme

The discretised form along the time axis of eqn (24) can be written as

$$c_{ij} \Delta u_i(\xi, T_N) - \int_{T_{N-1}}^{T_N} \int_S [G_{ij} \Delta t_i - F_{ij} \Delta u_i] \, dS \, d\tau - \int_{T_{N-1}}^{T_N} \int_V B_{ilj} \Delta \sigma_{il}^0 \, dV \, d\tau$$

$$= \int_{\tau=0}^{T_{N-1}} \int_S [G_{ij} \Delta t_i - F_{ij} \Delta u_i] \, dS \, d\tau + \int_{\tau=0}^{T_{N-1}} \int_V B_{ilj} \Delta \sigma_{il}^0 \, dV \, d\tau \qquad (30)$$

where the integral on the right-hand side is the contribution due to past dynamic history.

It is of interest that eqn (30), like eqn (24), still remains an exact formulation of the problem since no approximation has yet been introduced. However, in order to solve eqn (30), one has to approximate the time variation of the field quantities in addition to the usual approximation of spatial variation. For this purpose, the linear interpolation function in time described in the previous section is used.

Thus, after the usual discretisation and integrations (in both time and space), the integral equation (30) is transformed into an assembled system of equations of the form

$$[A_F^1]\{\Delta X^N\} - [B_F^1]\{\Delta Y^N\} - [C_F^1]\{(\Delta\sigma^0)^N\}$$

$$= - \sum_{n=2}^{N} [[A_F^n + A_I^{n-1}]\{\Delta X^{N-n+1}\} - [B_F^n + B_I^{n-1}]\{\Delta Y^{N-n+1}\}$$

$$+ [C_F^n + C_I^{n-1}]\{(\Delta\sigma^0)^{N-n+1}\}] \tag{31}$$

or

$$[A_F^1]\{\Delta X^N\} = [B_F^1]\{\Delta Y^N\} + [C_F^1]\{(\Delta\sigma^0)^N\} + \{R^N\} \tag{32}$$

or

$$A\Delta X^N = \Delta b^N + \Delta c^N \tag{33}$$

where

A and b are the matrices related to the unknown and known incremental displacements and tractions;

c is the matrix related to the initial stresses;

ΔX and ΔY are the vectors of unknown and known incremental displacements and tractions;

for $\Delta X, \Delta Y$ and $\Delta\sigma^0$, superscript denotes time, i.e. at time n, $\Delta X^n = X^n - X^{n-1}$;

for A, b and c matrices, the superscript denotes the time step when they are calculated, and the subscript denotes the local time node (I or F);

R^N is the effect of past dynamic history;

$$\Delta b^N = [B_F^1]\{\Delta Y^N\} + \{R^N\} \tag{34}$$

$$A = [A_F^1] \tag{35}$$

$$\Delta c^N = [C_F^1]\{(\Delta\sigma^0)^N\} \tag{36}$$

Similarly, the integral equation for stresses can be written in a discretised form as:

$$\{\Delta\sigma^N\} = [A_F^1]_\sigma\{\Delta X^N\} + [B_F^1]_\sigma\{\Delta Y^N\} + [C_F^1]_\sigma(\Delta\sigma^0)^N\} + \{R^N\}_\sigma \tag{37}$$

or

$$\Delta\sigma^N = A_\sigma\Delta X^N + \Delta b_\sigma^N + \Delta c_\sigma^N \tag{38}$$

where

$$A_\sigma = [A_F^1]_\sigma \tag{39}$$

$$\Delta b_\sigma^N = [B_F^1]_\sigma \{\Delta Y^N\} + \{R^N\}_\sigma \tag{40}$$

$$\Delta c_\sigma^T = [C_F^1]_\sigma \{(\Delta_\sigma^0)^N\} \tag{41}$$

The systems of equations (31) and (37) along with the constitutive equation (27) form the backbone of the inelastic transient dynamic analysis algorithm. For each time step essentially the static iterative algorithm has been followed (Banerjee and Raveendra, 1986). Further details can be found in Ahmad (1986) and Ahmad and Banerjee (1988c).

3.5 EXAMPLES OF APPLICATION

3.5.1 An Underground Explosion in a Layered Soil Medium

The surface vibration of a soil mass due to an underground explosion is investigated in this example. Sudden radial expansion of a cylindrical cavity is used to simulate the explosion. The soil is modelled by a viscoelastic soil layer overlying a viscoelastic half-space. The problem geometry is depicted in Fig. 1(a). Since the geometry and loading in the problem are symmetric, only one-half of the free surface, of the layer interface and of the cavity wall is modelled by boundary elements. A total of 34 boundary elements are used: 15 for the free surface, 15 for the layer interface and 4 to model the cavity wall. The material properties in consistent units of the top layer and the underlying half-space are assumed to be as follows, where μ = shear modulus, v = Poisson's ratio, ρ = material density and β = damping:

top layer: $\mu_1 = 647\,200$, $v_1 = 0.35$, $\rho_1 = 3.25$ and $\beta_1 = 0.05$

half-space: $\mu_2 = 1\,991\,150$, $v_2 = 0.3$, $\rho_2 = 2.85$ and $\beta_2 = 0.03$

The time history of the applied pressure on the cavity wall is shown in Fig. 1(b).

Figure 2(a) shows the time history of the vertical displacements at three selected points A(0, 0), B(2a, 0) and C(4a, 0) on the free surface, where a (= 10) is the radius of the cavity. It should be noted that the maximum vertical displacements at the free surface attenuate with distance. All displacements exhibit a sharp rise and then decay sharply without much oscillation, showing a high amount of geometric damping. The time histories of the horizontal displacements at points B and C are plotted in Fig. 2(b). The maximum magnitudes of the horizontal displacements are

P.K. BANERJEE, S. AHMAD AND H.C. WANG

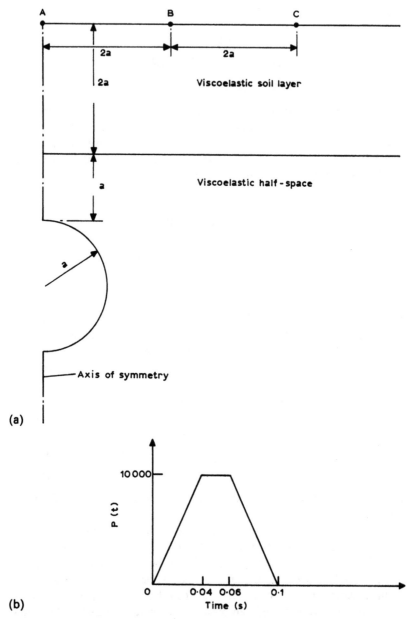

FIG. 1   (a) An underground explosion in a two-layered viscoelastic soil; (b) time history of the applied internal radial pressure on the cavity wall.

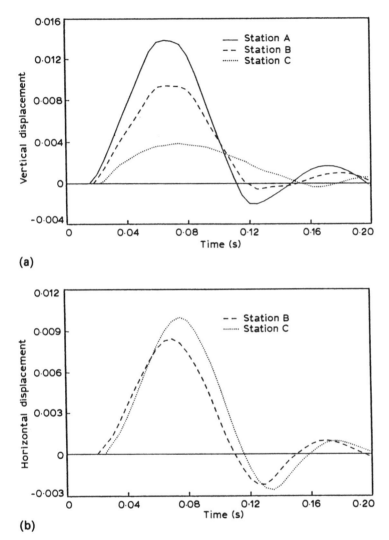

FIG. 2   (a) Time-history of vertical displacements; (b) time-history of horizontal displacements.

comparable to those of the vertical displacements at these points. This example shows that the present methodology is capable of solving even the complex problem of underground explosion in a multi-layered soil medium.

3.5.2 Application to the Vibration Isolation Problem

The present analysis has been applied to a number of vibration isolation examples and results are compared with the actual field observations.

Woods (1968) carried out a series of field tests in his attempt to define the screened zone and degree of amplitude reduction within the screened zone for trenches of a few specific shapes and sizes. He classified the foundation isolation problem into two categories, namely active isolation and passive isolation. Active isolation, as shown schematically in Fig. 3, is the

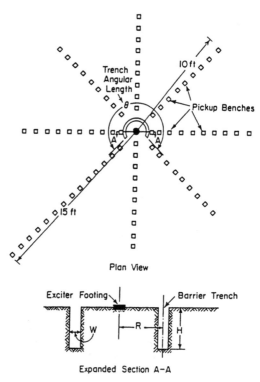

FIG. 3   Schematic of test layout for active isolation in the field (after Woods, 1968).

employment of barriers close to or surrounding the source of vibrations to reduce the amount of wave energy radiated away from the source. Passive isolation, as shown schematically in Fig. 4, is the employment of barriers at points remote from the source of vibrations but near a site where the amplitude of vibration must be reduced.

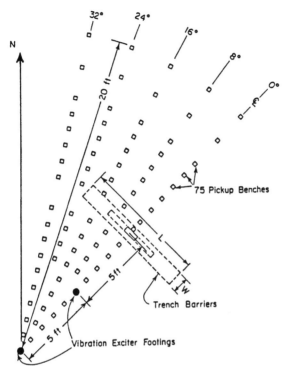

FIG. 4 Plan view of field site layout for screening at a distance (after Woods, 1968).

These tests were performed at a selected site of a two-layer system as shown in Fig. 5. A layer of uniform silty fine sand (SM) with dry density $\rho = 104 \, \text{lb/ft}^3$, water content $w = 7\%$, void ratio $e = 0.61$, pressure wave velocity $v_p = 940 \, \text{ft/s}$, and Rayleigh wave velocity $v_R = 413 \, \text{ft/s}$ rests on a layer of sandy silt (ML) with dry density $\rho = 91 \, \text{lb/ft}^3$, water content $w = 23\%$, void ratio $e = 0.68$, and pressure wave velocity $v_p = 1750 \, \text{ft/s}$ (Woods, 1968). The field tests for the active isolation were carried out for four lengths of trench aligned at 90, 180, 270, and 360°. Trench depths were 0·5, 1·0, and 2·0 ft, and trench width was 0·25 ft for all trenches. The distance from the centre of the footing to the centreline of the trench was 1·0 ft. A schematic diagram of this experimental setup is shown in Fig. 3. The field tests for the passive isolation were carried out for six lengths of trench of 1·0, 2·0, 3·0, 4·0, 6·0, and 8·0 ft. The distances from the centre of the footing to the centreline of the trench were 5 and 10 ft, and depths of trench were 1·0, 2·0, 3·0, and 4·0 ft. A schematic diagram of this experimental setup is shown in

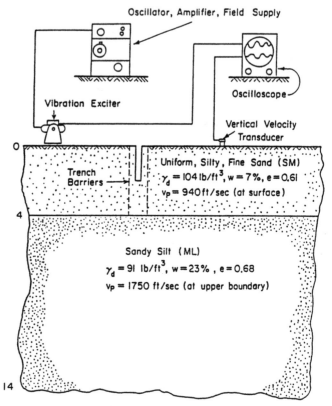

FIG. 5   Field site soil properties and schematic of instrumentation (after Woods, 1968).

Fig. 4. A constant input excitation force vector of 18 lb was used in all active and passive tests. The applied operational frequencies of the machine were 200, 250, 300, and 350 Hz.

In the present investigation, the soil profile is modelled as a two-layer system defined by the shear modulus G, Poisson's ratio v, and mass density ρ. The material properties of the top layer are basically determined by the relationships between the pressure wave velocity, the shear wave velocity, and the Rayleigh wave velocity. Accordingly, the values are taken as shear modulus $G = 647\,200\,\text{lb/ft}^2$, Poisson's ratio $v = 0.35$, and mass density $\rho = 3.25\,\text{lb s/ft}^3$. The material properties of the bottom layer are determined by assuming its Poisson's ratio to be the same as that of the top layer. Accordingly, the values are taken as shear modulus $G = 1\,991\,150\,\text{lb/ft}^2$,

Poisson's ratio $v = 0.35$, and mass density $\rho = 2.84\,\text{lb s/ft}^3$. The diameter of the footing of the vibration excitor, which was not specified, is assumed to be 0·5 ft. It was thought that this dimension has a relatively small influence on the results. A distributed pressure of $91.5\,\text{lb/ft}^2$, which is equivalent to 18 lb force, is applied to the footing.

The boundary element mesh used for the circular trench is shown in Fig. 6(a) and (b). Figure 6(a) shows the quadratic surface elements for the

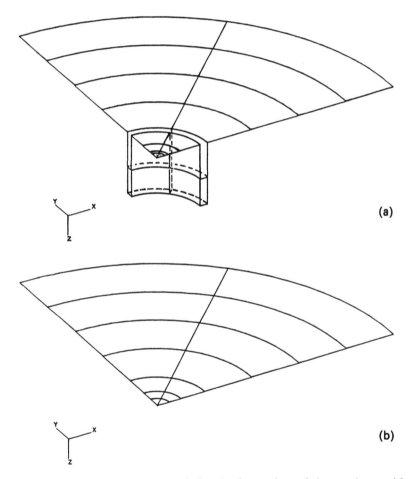

FIG. 6 (a) Boundary element mesh for the free-surface of the top layer with circular trench; (b) boundary element mesh for interface between the top layer and the half-space.

footing (shaded), trench and ground surface. Figure 6(b) shows the surface element pattern for the bottom surface of the top layer as well as for the top surface of the bottom layer. All elements used in these meshes were 6-noded triangles and 8-noded rectangles. The meshes used for the half-circle trench,

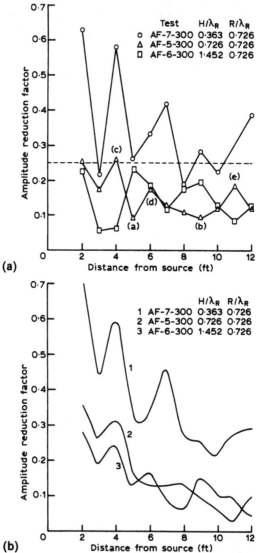

FIG. 7 Amplitude reduction factor versus distance from source: (a) field study (Woods, 1968); (b) BEM results.

the 270° circular trench and the rectangular trenches were essentially similar except for the details of the trench.

For the active isolation problem, comparisons of the BEM solutions and the experimental results are shown in Figs 7 to 10. Figure 7 shows the amplitude reduction factors for the full-circle trenches with three values of depth 0·5 ft (AF-7), 1·0 ft (AF-5), and 2·0 ft (AF-6) under an operating frequency of 300 Hz. It can be seen that the BEM solutions predict the amplitude reduction factor reasonably well. Figure 8 shows an overview of

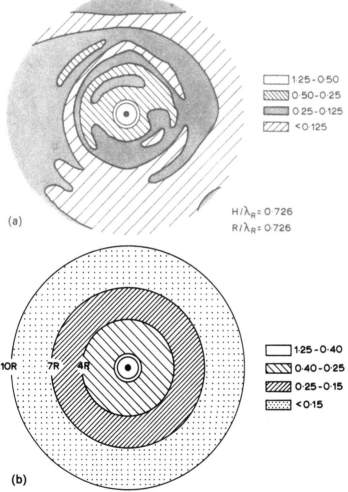

FIG. 8 Amplitude reduction factor contour diagram: (a) field study (Woods, 1968); (b) BEM results.

the contour diagram for case AF-5 at an operating frequency of 300 Hz. The lack of symmetry of the experimental results reflects the spatially random inhomogeneous characteristics of the soil. It can be seen that the BEM solution reasonably predicts the amplitude reduction factors. Figure 9 shows an overview of the contour diagram for a half-circle trench with

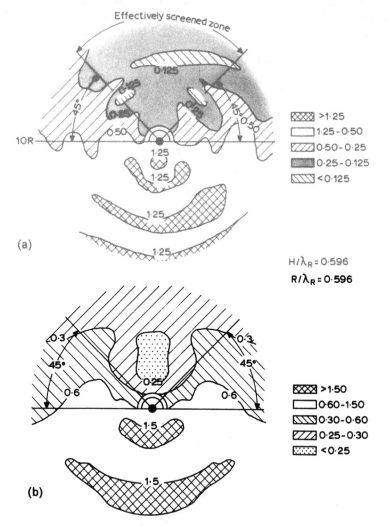

FIG. 9   Amplitude reduction factor contour diagram: (a) field study (Woods, 1968); (b) BEM results.

depth of 1·0 ft under an operating frequency of 250 Hz. Both approaches predict similar screened zones which are area-symmetrical about a radius from the source of excitation through the centre of the trench and bounded laterally by two radial lines extending from the centre of the source of excitation through points 45° from each end of the trench. Expansion of the screened zones may be noted when the screened zones in Fig. 10, which is obtained by extending the length of the trench from the half-circle to 270°, are compared with those of Fig. 9.

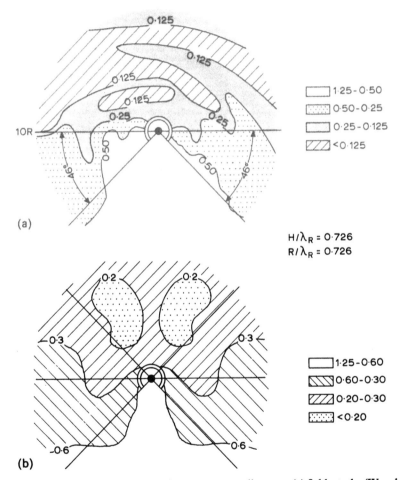

FIG. 10   Amplitude reduction factor contour diagram: (a) field study (Woods, 1968); (b) BEM results.

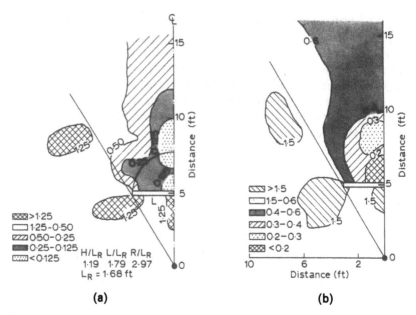

FIG. 11   Amplitude reduction factor contour diagram: (a) field study (Woods, 1968); (b) BEM results.

For the passive isolation problem, comparisons of the BEM solutions and the experimental results are shown in Figs 11 and 12. Figure 11 shows a contour diagram of the amplitude reduction factor for a rectangular trench 6 ft in length and 2 ft in depth under an operating frequency of 250 Hz. This trench is located 5 ft from the centre of the source. The comparison shows that again the experimental results are reasonably well predicted by the BEM solution. Figure 12 once again shows that the BEM solution predicts well the amplitude of the vertical displacement both in front of the trench and behind the trench.

It should be noted that the applications presented in this chapter represent some of the largest steady-state dynamic BEM analyses to date, which were carried out on a CRAY-XMP supercomputer. The computing times for the vibration isolation examples presented in Figs 9 to 15 were about 6 min for each problem.

3.5.3 A Flexible Square Plate Formulation on an Elastic Half-space

Karabalis and Beskos (1984), who described an approximate BEM implementation for half-space problems, examined the transient dynamic

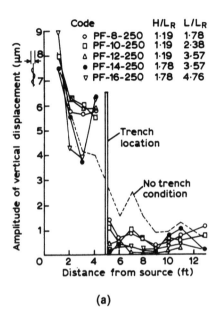

(a)

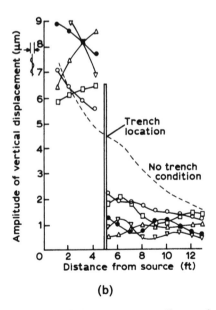

(b)

FIG. 12 Amplitude of vertical displacement versus distance from source: (a) field study (Woods, 1968); (b) BEM results.

response of a 60×60 in square, $11 \cdot 517$ in thick flexible foundation resting on the surface of an elastic half-space. The elastic constants for the plate were as follows: modulus of elasticity $E_p = 30 \cdot 004 \times 10^6$ psi, Poisson's ratio $\nu_p = 0 \cdot 3$, and mass density $\rho_p = 7 \cdot 34 \times 10^{-4}$ lb s^2/in^4. Those of the half-space were $E_s = 8 \cdot 84 \times 10^6$ psi, $\nu_s = 0 \cdot 3$, and $\rho_s = 2 \cdot 82 \times 10^{-4}$ lb s^2/in^4. This choice of elastic constants led Karabalis and Beskos to get a relative plate stiffness ratio $K = 0 \cdot 004$, where K is defined by

$$K = \frac{E_p h^3 (1 - \nu_s)}{12(1 - \nu_p^2)\mu_s b}$$

where h and b are the plate thickness and width, respectively, and μ_s is the shear modulus of the half-space.

Karabalis and Beskos used a finite element idealisation (4×4 mesh) for the plate and a boundary element idealisation (5×5 mesh) for the half-space. The boundary elements used were of constant variation over both space and time. This was obviously a very courageous attempt to solve such a difficult transient dynamic problem.

In the present attempt at analysing the same problem, both the plate and the surface of the half-space were modelled by BEM. The meshes used for one-quarter of the plate region are shown in Fig. 13(a) for the coarse mesh and in Fig. 14(a) for the fine mesh. Similarly two sets of meshes (coarse and fine) for the half-space region are shown in Figs 13(b) and 14(b). In all cases, quadratic elements with linear time variations were used.

The results of the present analysis for a dynamic pressure pulse shown in Fig. 15(a) are plotted in Fig. 15(b) where the results of Karabalis and Beskos are also shown for comparison. It should be noted that in the present model a plate of finite thickness has resulted in a slight delay in the wave front reaching the half-space. This, however, does not account for the very large discrepancies in the response times.

3.5.4 Vibration of a Rigid Circular Disc

The generalised axisymmetric formulation developed for the periodic dynamic analysis has been implemented in GPBEST (which additionally can analyse axisymmetric transient dynamic problems). As an illustrative example, the problem of vertical and rocking vibrations of a rigid circular disc resting on top of an elastic half-space is examined in Figs 16 and 17. A total of four quadratic elements over the radius and 18 quadratic elements outside the loaded area of up to a distance of 12 times the radius

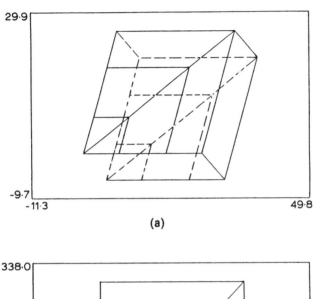

(a)

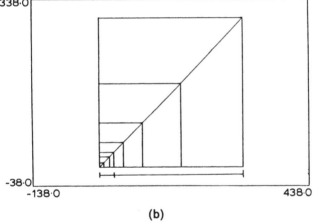

(b)

FIG. 13   Boundary Element Model for a flexible square footing on an elastic half-space (mesh #1): (a) footing mesh; (b) half-space mesh.

were used to obtain these results. Figures 16 and 17 respectively show the non-dimensional vertical stiffness and rocking stiffness plotted against the non-dimensional frequency along with those obtained earlier analytically by Veletsos and Wei (1971). Results for vertical stiffness are in excellent agreement with the analytical results, but there is some difference between the results for the rocking response.

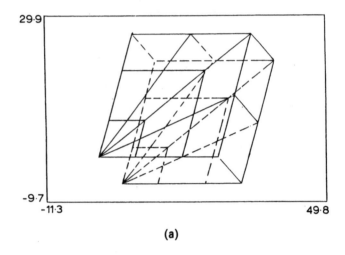

(a)

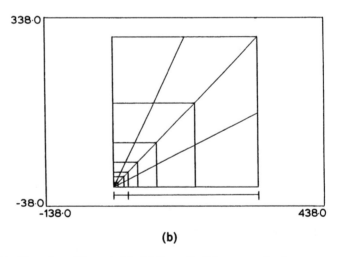

(b)

FIG. 14 Boundary Element Model for a flexible square footing on an elastic half-space (mesh #2): (a) footing mesh; (b) half-space mesh.

3.5.5 A Buried Caisson Foundation

Figure 18 shows a typical caisson foundation with a depth of 14 m and a radius of 4 m. The thickness of the side wall is 0·6 m and the cap and bottom seal are each 1 m thick. The material properties of the caisson and

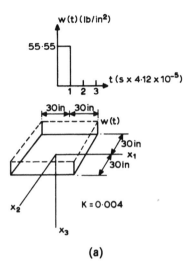

(a)

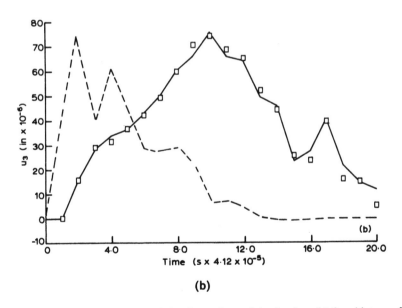

(b)

FIG. 15   (a) Loading curve and the dimensions of the footing; (b) time history of vertical displacement at the centre of the footing: —— present analysis (mesh 1); □—□ present analysis (mesh 2); — — — Karabalis and Beskos (1984).

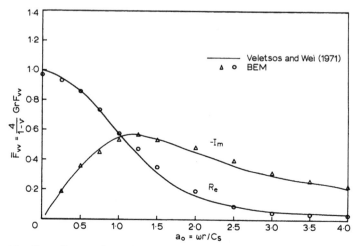

FIG. 16   Compliance of a rigid massless circular plate on an elastic half-space ($v = 1/3$): vertical response.

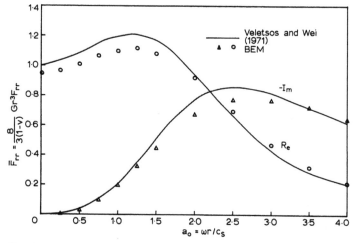

FIG. 17   Compliance of a rigid massless circular plate on an elastic half-space ($v = 1/3$): rocking response.

refilled materials, compared to the surrounding soil, are

Caisson material: $E_c/E_s = 500$, $\rho_c/\rho_s = 1\cdot5$, $\beta_f/\beta_s = 0$

Refilled material: $E_f/E_s = 10$, $\rho_f/\rho_s = 1\cdot2$, $\beta_f/\beta_s = 1$

where E, ρ, β are the Young's modulus, mass density and damping of

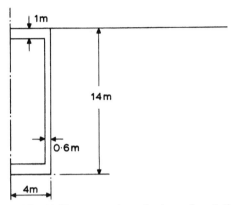

FIG. 18   Half cross-section of caisson foundation.

materials. The damping ratio of soil is taken as 0·02 and the Poisson's ratios for all materials are assumed to be 0·25.

This problem was recently examined by Chen and Penzien (1986), using a combination of the finite element method and a symmetric form of the indirect boundary element method.

In order to apply the present analysis to this problem, the internal caisson boundary, the external caisson boundary and the soil boundary surfaces were modelled using 16, 16 and 24 quadratic boundary elements respectively.

Figure 19 shows the results of the present analysis for the vertical

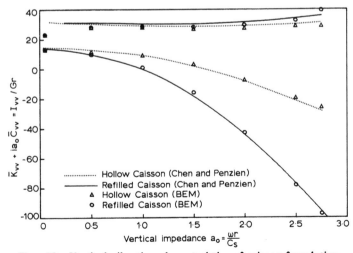

FIG. 19   Vertical vibration characteristics of caisson foundation.

vibration of the caisson plotted against the non-dimensional frequency, together with the results of Chen and Penzien (1986). Both sets of results for the cases of hollow caisson and refilled caisson agree well with each other.

3.5.6 Surface Response of an Elastic Half-space Due to Uniform Initial Strain in a Subsurface Cuboidal Region

As shown in Fig. 20(a) a cuboidal region of dimensions $2b \times 2b \times 2b$, embedded at a depth D within a semi-infinite solid, is subjected to a uniform

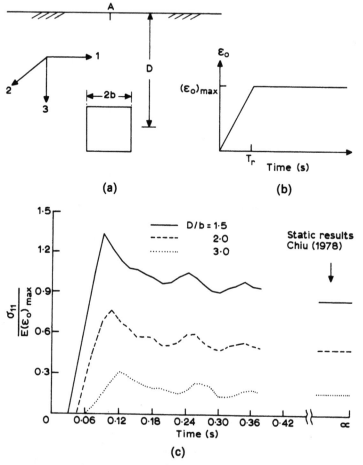

FIG. 20 (a) A cube embedded in an elastic half-space; (b) time history of applied initial strains; (c) time history of horizontal stresses $(\sigma_{11} = \sigma_{22})$ $(T_r = 0.06 \, \text{s}; \, \varepsilon_{11} = \varepsilon_{22} = \varepsilon_{33} = \varepsilon_0)$.

initial strain. This problem was first examined by Chiu (1978) under a static initial strain system. He was able to obtain closed-form solutions for surface deformation and stresses.

Essentially the same problem has been examined here under a dynamic initial strain disturbance such as is likely to occur in a solid due to dynamic inelastic strain development. The time history of the applied dynamic initial strain is shown in Fig. 20(b), where T_r denotes the time for the initial strain to rise to its maximum value. The purpose of choosing this type of time history was to ensure that the response asymptotically approaches its static value at later times. The necessary details of geometry and material properties are $b = 1.0$, $E = 1000$, $v = 0.3$ and $\rho = 1.0$. The depth of embedment D and rise time T_r were assigned different values ($D/b = 1.5, 2.0$ and 3.0; $T_r = 0.03$ and 0.06 s).

For the BEM analysis of this problem, the ground surface was modelled by 16 six- and eight-noded quadrilateral surface elements and the cuboidal region was modelled by a 20-noded volume cell. The time step used in this example is $\Delta T = 0.015$ s.

Figure 20(c) shows the effect of the depth of embedment (D/b) on the horizontal stresses ($\sigma_{11} = \sigma_{22}$) at point A (Fig. 16(a)) on the free surface due to volumetric initial strain σ_0 with rise time $T_r = 0.06$ s. The stress is plotted as a non-dimensional quantity $\sigma_{11}/[E(\varepsilon_0)_{max}]$. It can be seen that the results for all three depths of embedment first rise to a peak value and then at longer times asymptotically approach Chiu's exact static solution. The dynamic magnification factors for $D/b = 1.5, 2.0$ and 3.0 and $1.59, 1.58$ and 1.95, respectively. The dynamic magnification factor is the ratio between the peak value and the static value. It can also be seen from Fig. 16(c) that the time for reaching the peak value decreases with decreasing depth of embedment.

Figure 21(a) shows essentially similar results for vertical ground displacements at point A, normalised as $u_3/[b(\varepsilon_0)_{max}]$. As before, it can be seen that the results from the present analysis asymptotically approach the static solution of Chiu at later times. From Fig. 17(a), it is also evident that the magnification factors for vertical displacements are smaller compared to those of horizontal stresses. However, for a given depth of embedment (D/b), both the vertical displacement and the horizontal stress reach their respective peak values at the same time (e.g. for $D/b = 1.5$, time for maximum response is 0.09 s).

Finally, Fig. 21(b) compares the vertical displacements at point A due to three different initial strains (one volumetric and two direct strains) considered separately for a rise time $T_r = 0.03$ s and a depth of embedment $D/b = 1.5$. Both the volumetric and vertical initial strains lead to higher

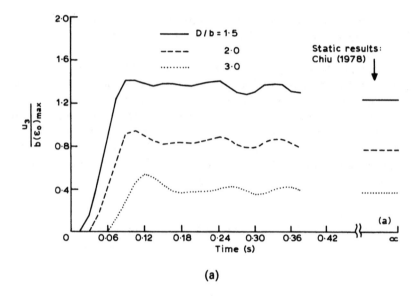

(a)

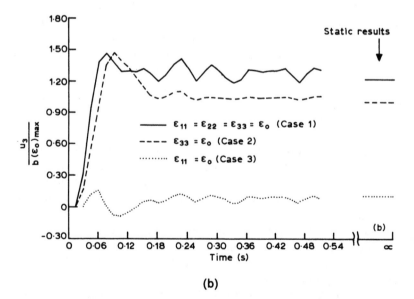

(b)

FIG. 21   (a) Time history of vertical displacements ($\varepsilon_{11} = \varepsilon_{22} = \varepsilon_{33} = \varepsilon_0$; $T_r = 0.06$ s);
(b) time history of vertical displacements ($D/b = 1.5$; $T_r = 0.03$ s).

displacements. Once again all the dynamic solutions finally decay to their respective static values at longer times. Other examples of inelastic dynamic analysis can be found in Ahmad (1986) and Ahmad and Banerjee (1988c).

3.6 CONCLUSIONS

A very sophisticated implementation of the BEM formulation for any two-dimensional, axisymmetric and three-dimensional problems of steady-state and transient dynamics has been presented. The analysis uses curvilinear boundary elements, self-adaptive error control in the integration, and a completely general multi-region assembly for multiple substructured regions. The developed analysis has been applied to a number of complex problems to demonstrate that it can be used to carry out an efficient dynamic analysis of three-dimensional problems.

The transient formulation has been extended to non-linear dynamics. In the present work, the Von Mises constitutive relationship is used to model the material behaviour. However, for materials like soils, a more realistic material model needs to be included in the present code. Once the new material models are included, the proposed methodology for time-domain non-linear transient dynamic analysis has the potential to provide a valuable tool for solving soil-structure interaction problems, which cannot be accomplished by the available frequency-domain algorithms.

ACKNOWLEDGEMENTS

The work presented in this chapter was made possible by NASA Contract NAS3-23697 and a grant from United Technologies Corporation. The authors are indebted to Dr C. Chamis, the NASA Project Manager, and Dr E. Todd, the Pratt & Whitney Project Manager, for their support and encouragement, and to Dr R.B. Wilson and N. Miller of Pratt & Whitney for valuable discussions.

REFERENCES

AHMAD, S. (1986). Linear and nonlinear dynamic analysis by boundary element method, Ph.D. Thesis, State University of New York at Buffalo.
AHMAD, S. & BANERJEE, P.K (1988a). Multi-domain BEM for two-dimensional problems of elastodynamics, *Int. J. Numer. Methods Eng.*, **26**, 891–911.

AHMAD, S. & BANERJEE, P.K. (1988*b*). Time domain transient elastodynamic analysis of 3-D solids by BEM, *Int. J. Numer. Methods Eng.*, **26**, 1709–28.

AHMAD, S. & BANERJEE, P.K. (1988*c*). Development of BEM for dynamic plasticity, to appear in *Int. J. Numer. Methods Eng.*

AHMAD, S. & MANOLIS, G.D. (1987). Dynamic analysis of 3-D structures by a transformed boundary element method, *Comput. Mech.*, **2**, 185–96.

BANERJEE, P.K. & AHMAD, S. (1985). Advanced three-dimensional dynamic analysis of 3-D problems by boundary element methods, *Proc. ASME Conf. on Advanced Topics in Boundary Element Analysis*, Eds T.A. Cruse, A.B. Pifko and H. Armen, AMD Vol. 72, ASME, New York, 65–81.

BANERJEE, P.K. & BUTTERFIELD, R. (1981). *Boundary Element Methods in Engineering Science*, McGraw-Hill, London. Also US Edition, New York, 1983; Russian Edition, Moscow, 1984.

BANERJEE, P.K. & DAVIES, T.G. (1984). Advanced implementation of boundary element methods for three-dimensional problems of elastoplasticity and viscoplasticity. Chapter 1 in: *Developments in Boundary Element Methods—3*, Eds P.K. Banerjee and S. Mukherjee, Elsevier Applied Science Publishers, London, 1–26.

BANERJEE, P.K. & RAVEENDRA, S.T. (1986). Advanced boundary element analysis of two- and three-dimensional problems of elastoplasticity, *Int. J. Numer. Methods Eng.*, **23**, 985–1002.

BANERJEE, P.K., AHMAD, S. & MANOLIS, G.D. (1986). Transient elastodynamic analysis of three-dimensional problems by boundary element method, *Earthquake Eng. Structural Dyn.*, **14**, 933–49.

BANERJEE, P.K., AHMAD, S. & CHEN, K.H. (1988). Advanced applications of BEM to wave barriers in multi-layered soil media, *Earthquake Eng. Structural Dyn.*, **16**, 1041–60.

CHEN, C.H. & PENZIEN, J. (1986). Dynamic modelling of axisymmetric foundations. *Earthquake Eng. Structural Dyn.*, **14**, 823–40.

CHIU, Y.P. (1978). On the stress field and surface deformation in a half space with a cuboidal zone in which initial strains are uniform, *J. Appl. Mech., Trans. ASME*, **45**, 302–6.

GRAFFI, D. (1947). Memorie della Academia Scienze, *Bologna Series 10*, **4**, 103–9.

HENRY, D.P., Jr (1987). Advanced development of the boundary element method for elastic and inelastic thermal stress analysis, Ph.D. Thesis, State University of New York at Buffalo.

HILL, R. (1950). *The Mathematical Theory of Plasticity*, Clarendon Press, Oxford, UK.

KARABALIS, D.L. & BESKOS, D.E. (1984). *Dynamic Response of 3-D Foundations by Time Domain Boundary Element Method*, Final Report Part A, NSF Grant No. CEE-8024725, Department of Civil and Mineral Engineering, University of Minnesota, Minneapolis, Minnesota.

MANOLIS, G.D. & BESKOS, D. (1986). Private communication with a copy of Ph.D. Thesis of S. Ahmad (1986), University of Patras, Greece, June 1986.

MUKHERJEE, S. (1982). *Boundary Elements in Creep and Fracture*, Applied Science Publishers, London.

RIZZO, F.J. & SHIPPY, D. (1979). A boundary integral approach to potential and elasticity problems for axisymmetric bodies with arbitrary boundary conditions, *Mech. Res. Commun.*, **6**, 99–103.

VELETSOS, A.S. & WEI, Y.T. (1971). Lateral and rocking vibration of footings, *J. Soil Mech. Foundation Eng., Am. Soc. Civ. Eng.*, **97**, 1227–48.

WILSON, R.B. & BANERJEE, P.K. (1986). *3-D Inelastic Analysis Methods for Hot Section Components*, Third Annual Status Report, Vol. II, NASA Contract Report NAS 179517.

WOODS, R.D. (1968). Screening of surface waves in soils, *Proc. Am. Soc. Civ. Eng.*, **94**, SM4, 951–79.

Chapter 4

BOUNDARY ELEMENT METHODS FOR POROELASTIC AND THERMOELASTIC ANALYSES

G.F. DARGUSH and P.K. BANERJEE

Department of Civil Engineering,
State University of New York at Buffalo, USA

SUMMARY

The development of a general boundary element method (BEM) for two- and three-dimensional quasistatic poroelasticity is discussed in detail. The new formulation, for the complete Biot consolidation theory, operates directly in the time domain and requires only boundary discretisaton. Consequently, the dimensionality of the problem is reduced by one and the method becomes quite attractive for geotechnical analyses, particularly those which involve extensive or infinite domains. Furthermore, as a result of a well-known analogy, the new BEM formulation is equally applicable for quasistatic thermoelasticity.

The presentation includes the definition of the two key ingredients for the BEM, namely the fundamental solutions and a reciprocal theorem. These formulations are implemented in a general purpose BEM system which includes higher-order conforming elements, self-adaptive integration, and multi-region capability. A number of detailed examples are presented to illustrate the accuracy and suitability of this boundary element approach for both consolidation and thermomechanical analyses.

4.1. INTRODUCTION

This chapter addresses problems of quasistatic poroelasticity, and, via the well-established analogy, thermoelasticity. A boundary-only, time-domain

BEM formulation and implementation is detailed for both two- and three-dimensional bodies. This new BEM is particularly advantageous for poroelasticity since it can readily handle the infinite or semi-infinite domains that are often associated with geotechnical problems. On the other hand, for thermoelasticity, this BEM formulation has the ability to capture the steep thermal gradients and stresses that may occur near the surface during severe transients. Additionally, the time-domain formulation eliminates the need for numerical transform inversions and permits the extension to non-linear phenomena.

Over the past dozen years, considerable attention has been directed towards these problems. The earliest work, done by Rizzo and Shippy (1977) and Cruse et al. (1977), pertained to steady-state thermoelasticity. The approach consisted of an initial phase in which a boundary element analysis of steady-state heat conduction was employed to determine the surface temperature and flux distribution. In the second phase, the resulting temperatures were applied as body forces in an elastostatic BEM to obtain deformation and stress. General properties of the steady-state temperature field were exploited to reduce the thermal body force volume integral to surface integrals involving the known boundary temperatures and flux. Thus, the entire two-step process required only surface discretisation.

In the quasistatic realm, Banerjee and Butterfield (1981, 1982) presented a staggered procedure for solving the coupled equations. The algorithm requires the solutions of the transient pore fluid (or heat) flow equation followed by an elastic analysis including body forces at each time step. Unfortunately, this is not a boundary-only formulation, and complete volume discretisation is required. As an example, the consolidation of a strip foundation was examined. Interestingly, this same scheme was applied by Kuroki et al. (1982) for two-dimensional consolidation, by Aramaki and Yasuhara (1985) for axisymmetric consolidation, and by Aramaki (1986) for consolidation problems which include thin layers with high permeability.

Another alternative involves casting the problem in the Laplace transform domain. Cheng and Liggett (1984) took that approach for two-dimensional quasistatic poroelasticity. While this is a boundary-only formulation, the procedure is not, in general, very satisfactory. Transformation to the Laplace domain introduces several serious side-effects, including sensitivity to the selection of values for the transform parameter, requirements for numerical inversion of the transform, and limitations to strictly linear problems. It should be noted that in an earlier paper Cleary (1977) had discussed some aspects of a similar numerical procedure.

More recently, Nishimura (1985) developed the first two-dimensional time-domain boundary element formulation for regions infiltrated by an incompressible fluid. The fundamental solutions were presented, along with discussions of existing singularities. Numerical results for a simple, one-element problem were included. In a subsequent paper, Nishimura *et al.* (1986) presented two additional problems, but provided no further correlations with analytical results.

Returning to thermoelasticity, Tanaka and Tanaka (1981) presented a reciprocal theorem and the corresponding boundary element formulation for the time-domain coupled problem. However, kernel functions are not discussed and no numerical results are included. In fact, later, Tanaka *et al.* (1984) chose instead to implement the volume-based thermal body force approach of Banerjee and Butterfield. In a series of papers, Sládek and Sládek (1983, 1984*a, b*) presented a collection of fundamental solutions, in both Laplace transform and time domains, under the classifications of coupled, uncoupled, transient, and quasistatic thermoelasticity. Boundary integral equations were also included, although in several instances these were written inappropriately in terms of displacement and traction rates. Kernel singularities were not discussed and a numerical implementation was not attempted. More recently, Chaudouet (1987) again resorts to the volume-based approach, while Masinda (1984) and Sharp and Crouch (1986) have directed efforts towards converting the thermal body force volume integral into a surface integral. Masinda, working in three dimensions, presents some formulations, but stops short before attempting an implementation. On the other hand, Sharp and Crouch develop an approach for two-dimensional quasistatic thermoelasticity using time-dependent Green's functions. However, the authors then introduce volume integrals in their time-marching algorithm. This, unfortunately, undermines most of the advantages of the BEM.

The present effort involves the development of boundary element methods for general two- and three-dimensional Biot's theory of consolidation and the corresponding theory of coupled thermoelasticity (Dargush, 1987; Banerjee and Dargush, 1988; Dargush and Banerjee, 1989*a, b, c*). Included is the capability to analyse arbitrarily shaped, multi-region bodies subjected to time-varying boundary conditions. Furthermore, relationships are developed to determine accurately the effective stress in the interior, as well as on the boundary. The resulting BEM time-domain formulation and implementation represents the first of its kind for three-dimensional problems. However, even in the two-dimensional case, the present formulation appears to be more attractive than that of Nishimura.

In particular, Nishimura introduces the time-integrated total normal mass flow as a primary variable in order to avoid a singularity problem. Unfortunately, this choice is not consistent with the remaining primary variables which are intensive, rather than extensive in nature. Beyond this, proper boundary conditions for the poroelastic problem are in terms of flux, not the total flow. Once the fundamental solutions are clearly understood, all the singularities become manageable, and flux can be used directly as the primary variable.

Additionally, Nishimura omits the term involving the time rate of change of pore pressure from the flow equation, thus limiting the formulation to completely saturated soils. The more general Biot's theory, employed in the present work, is also applicable for partially saturated soils, and for fluid-infiltrated rocks for which compressibility of the solid phase becomes important. Additionally, the thermoelastic analogy is only valid for the complete Biot theory.

Before presenting the detailed development of the new boundary element method, some background information is needed. Consequently, the governing equations and fundamental solutions are provided in the next two sections. Indicial notation is used throughout. Thus, summations are implied by repeated indices, commas represent differentiation with respect to spatial coordinates, and a superposed dot denotes a material time derivative.

4.2 GOVERNING EQUATIONS

4.2.1 Poroelasticity

In the absence of inertial effects, application of the conservation of momentum to a poroelastic body yields the familiar equilibrium equations,

$$\sigma_{ji,j} + f_i = 0 \tag{1}$$

in which σ_{ji} represents the total stress tensor. Meanwhile from kinematics, the linearised strain–displacement relationship takes the form

$$\varepsilon_{ij} = \tfrac{1}{2}(u_{i,j} + u_{j,i}) \tag{2}$$

Next, a constitutive law for the soil mass must be introduced. Following Biot (1941), the total stress σ_{ij} is decomposed into effective stress σ'_{ij} of the soil and pore pressure p of the fluid. That is,

$$\sigma_{ij} = \sigma'_{ij} - \delta_{ij}\beta p \tag{3}$$

Note that a negative sign appears because, by convention, the stresses σ_{ij} and σ'_{ij} are positive in tension, whereas positive pore pressure is compressive.

Then the effective stress is related to the strains via an isotropic, linear elastic constitutive law of the form

$$\sigma'_{ij} = \delta_{ij}\lambda\varepsilon_{kk} + 2\mu\varepsilon_{ij} \tag{4}$$

The elastic constants, λ and μ, appearing in eqn (4) are those obtained from soil samples under drained conditions. Consequently, λ and μ are termed drained moduli. In contrast, material properties measured at very short time intervals after load application, in order to prevent diffusion of the fluid, are labelled undrained. With that in mind, the dimensionless material parameter in eqn (3) can be defined as

$$\beta = 1 - \frac{K}{K'_s} \tag{5}$$

where

$$K = \lambda + \frac{2\mu}{3} = \text{drained bulk modulus}$$

and K'_s is an empirical constant which in certain circumstances equals the bulk modulus of the solid constituents (Rice and Cleary, 1976). For soils in which both the fluid and solid constituents are incompressible, $\beta = 1$. Equation (3) then reduces to Terzaghi's definition of effective stress.

Now, eqns (1)–(4) can be combined to produce Navier's equations generalised for a poroelastic body. That is,

$$(\lambda + \mu)u_{j,ij} + \mu u_{i,jj} - \beta p_{,i} + f_i = 0 \tag{6}$$

Note that eqn (6) involves the pore pressure of the fluid, as well as the displacement of the soil.

The set of governing differential equations for poroelasticity can be completed by applying the principle of conservation of mass to the fluid. In differential form, this is written as

$$q_{i,i} + \dot{m} = \bar{\psi} \tag{7}$$

where

q_i = fluid mass flux vector
m = fluid mass content per unit volume
$\bar{\psi}$ = rate of fluid mass supply per unit volume.

Next, Darcy's law is invoked to relate flux to pore pressure gradient in a linear fashion. For an isotropic medium this becomes simply

$$q_i = \rho_0 \kappa p_{,i} \tag{8}$$

with fluid mass density ρ_0 and permeability κ. The fluid mass content must also be expressed in terms of displacements and pore pressure. A convenient form of this relationship, given by Rudnicki (1987), is

$$m - m_0 = \beta \rho_0 \left(u_{j,j} + \frac{\beta p}{\lambda_u - \lambda} \right) \tag{9}$$

in which m_0 is the reference value of m, and λ_u is the undrained Làme modulus. Note that eqn (9) could instead be written in terms of the undrained Poisson's ratio, ν_u, where

$$\nu_u = \frac{\lambda_u}{2(\lambda_u + \mu)} \tag{10}$$

(In the special case of incompressible constituents, $\nu_u \to 0.5$ and $\lambda_u \to \infty$.) Introducing eqns (8) and (9) into eqn (7) leads to the following expression for mass conservation:

$$\rho_0 \kappa p_{,jj} - \frac{\rho_0 \beta^2}{\lambda_u - \lambda} \dot{p} - \rho_0 \beta \dot{u}_{j,j} + \bar{\psi} = 0 \tag{11}$$

After letting $\bar{\psi} = \rho_0 \psi$ in eqn (11), the complete set of governing differential equations for poroelasticity can be rewritten as

$$(\lambda + \mu) u_{j,ij} + \mu u_{i,jj} - \beta p_{,i} + f_i = 0 \tag{12a}$$

$$\kappa p_{,jj} - \frac{\beta^2}{\lambda_u - \lambda} \dot{p} - \beta \dot{u}_{j,j} + \psi = 0 \tag{12b}$$

Note, in particular, the appearance of displacement and pore pressure in both equations. Thus, in poroelasticity not only do pore pressure changes cause deformation, but also deformation produces pore pressure variations. Clearly, the set of equations (12) must be solved simultaneously.

The material parameters present in eqns (12) are not necessarily those typically measured for soils, and other choices (e.g. Skempton's pore pressure coefficient B) are perfectly valid. Table 1 details the interrelationship between some of the more useful parameters.

In addition to eqns (12), for a well-posed problem, boundary and initial conditions must be specified. Formally, the boundary conditions for all

TABLE 1
SELECTED POROELASTIC CONSTANTS WITH INTERRELATIONSHIPS

Definitions

B	Skempton's pore pressure coefficient
E	Drained elastic modulus
G	Shear modulus
K	Drained bulk modulus
K'_s	Empirical constant in units of K
K_u	Undrained bulk modulus
β	Non-dimensional parameter
μ, λ	Drained Lame moduli
μ_u, λ_u	Undrained Lame moduli
v	Drained Poisson's ratio
v_u	Undrained Poisson's ratio

Relationships

$$\mu = G = \frac{E}{2(1+v)} \qquad \lambda = \frac{Ev}{(1+v)(1-2v)}$$

$$\mu_u = \mu \qquad \lambda_u = \frac{Ev_u}{(1+v)(1-2v_u)}$$

$$v_u = \frac{\lambda_u}{2(\lambda_u + \mu)} \qquad K = \lambda + \frac{2\mu}{3} \quad K_u = \lambda_u + \frac{2\mu}{3}$$

$$\beta = 1 - \frac{K}{K'_s} = \frac{1}{B}\left(1 - \frac{K}{K_u}\right) = \frac{3}{B}\left(\frac{\lambda_u - \lambda}{3\lambda_u + 2\mu}\right) = \frac{3}{B}\left[\frac{v_u - v}{(1+v_u)(1-2v)}\right]$$

$$K'_s = \frac{BKK_u}{K+(B-1)K_u} \qquad K_u = \frac{K}{1-B\beta} \qquad \lambda_u = \lambda + \frac{B\beta K}{1-B\beta}$$

points X in S are given by

$$u_i = U_i(X,t) \tag{13a}$$
or
$$t_i = T_i(X,t) \tag{13b}$$
or
$$t_i = K(X,t)u_i \tag{13c}$$
and
$$p = P(X,t) \tag{14a}$$
or
$$q = Q(X,t) \tag{14b}$$
or
$$q = H(X,t)p \tag{14c}$$

while the initial conditions

$$u_i = U_i^0(Z) \tag{15a}$$

$$p = P^0(Z) \tag{15b}$$

are required for all points Z in V at time zero. In the above, q is the flux normal to the surface S, t_i is the traction vector, $K(X, t)$ is a spring stiffness, and $H(X, t)$ is a smearing contact resistance. The specification of eqns (13), (14) and (15) along with eqns (12) completely define the poroelastic problem.

An interesting reformulation of eqns (12) in terms of the fluid mass content leads to the following set of governing differential equations:

$$(\lambda_u + \mu)u_{j,ij} + \mu u_{i,jj} - \left(\frac{\lambda_u - \lambda}{\beta}\right)m_{,i} = 0 \tag{16a}$$

$$\kappa\left(\frac{\lambda_u - \lambda}{\beta^2}\right)\left(\frac{\lambda + 2\mu}{\lambda_u + 2\mu}\right)m_{,jj} = \dot{m} \tag{16b}$$

Now, upon examination of eqn (16b), it is evident that in poroelasticity the fluid mass content must satisfy a simple diffusion equation. The diffusivity, c, is strictly a material characteristic, which in this case becomes

$$c = \kappa\left(\frac{\lambda_u - \lambda}{\beta^2}\right)\left(\frac{\lambda + 2\mu}{\lambda_u + 2\mu}\right) \tag{17}$$

4.2.2 Thermoelasticity

A complete derivation of thermoelastic theory can be found in the textbooks by Boley and Weiner (1960) and Nowacki (1986). Consequently, only a few of the key assumptions will be mentioned before presenting the governing differential equations. In particular, the classical theory assumes infinitesimal deformations and linear isotropic materials. Thus, the linearised strain–displacement relationship

$$\varepsilon_{ij} = \tfrac{1}{2}(u_{i,j} + u_{j,i}) \tag{18}$$

is once again employed, along with the constitutive laws of Duhamel–Neumann,

$$\sigma_{ij} = \delta_{ij}\lambda\varepsilon_{kk} + 2\mu\varepsilon_{ij} - \delta_{ij}(3\lambda + 2\mu)\alpha(\theta - \theta_0) \tag{19}$$

and Fourier,

$$q_i = -k\theta_{,i} \tag{20}$$

In eqns (19) and (20), θ is the present temperature, while θ_0 represents the temperature of a zero stress reference state, and q_i is the heat flux vector. Additionally, λ and μ are the usual Lame's isothermal elastic constants, α is the coefficient of thermal expansion, and k represents the isotropic thermal conductivity.

Then, application of the laws of conservation of momentum and energy lead, in the absence of inertia, to the following set of equations:

$$(\lambda+\mu)u_{j,ij}+\mu u_{i,jj}-(3\lambda+2\mu)\alpha\theta_{,i}+f_i=0 \qquad (21a)$$

$$k\theta_{,jj}-\rho c_\varepsilon\dot{\theta}-(3\lambda+2\mu)\alpha\theta_0\dot{u}_{j,j}+\psi=0 \qquad (21b)$$

in which ρ is the mass density, c_ε is the specific heat at constant deformation, while f_i and ψ are the body forces and sources, respectively. The theory portrayed by eqns (21) is formally named Coupled Quasistatic Thermoelasticity (CQT). Notice, in particular, the appearance of displacement and temperature in both equations. Thus, in CQT not only do changes in temperature cause deformation, but also deformation produces temperature variations. In general, the set of equations (21) must be solved simultaneously. However, from a practical engineering standpoint, it has been determined by a number of researchers (e.g. Boley and Weiner, 1960; Day, 1982) that the term involving displacement in eqn (21b) is negligible, thus uncoupling the momentum and energy balance equations. Dropping that term, eqn (21b) can first be solved independently for the temperature field. Subsequently, displacements are determined from eqn (21a) with the known temperature distribution. This is, in fact, the accepted procedure for thermal stress analysis. In the present work the general, fully coupled theory governed by eqns (21) will be retained. The more efficient uncoupled approach is detailed in Dargush (1987) and Dargush and Banerjee (1988b, c).

To complete the formulation of a well-posed problem, boundary and initial conditions must, of course, be specified. Formally, the boundary conditions for all points X in S can be written as

$$u_i=U_i(X,t) \qquad (22a)$$

or

$$t_i=T_i(X,t) \qquad (22b)$$

or

$$t_i=K(X,t)u_i \qquad (22c)$$

and

$$\theta=\Theta(X,t) \qquad (23a)$$

or

$$q=Q(X,t) \qquad (23b)$$

or

$$q=H(X,t)\,[\Theta_{amb}(X,t)-\theta] \qquad (23c)$$

In addition, the initial conditions

$$u_i = U_i^0(Z) \tag{24a}$$

$$\theta = \Theta^0(Z) \tag{24b}$$

are required for all points Z in V at time zero which in the present analysis has been assumed to be zero for simplicity. In the above, q is the heat flux normal to the surface S, and t_i is the traction vector defined by

$$t_i = \sigma_{ji} n_j \tag{25}$$

Note that eqn (23c) represents the familiar convection boundary condition, in which $H(X, t)$ and $\Theta_{amb}(X, t)$ denote the film coefficient and ambient fluid temperature, respectively. The specification of eqns (22), (23) and (24) along with eqns (21) completely defines the CQT problem.

However, the governing differential equations (21) can be reformulated by employing the entropy density, s, rather than the temperature as an explicit variable. The result is

$$\left(\lambda + \frac{\theta_0 \beta^2}{\rho c_\varepsilon} + \mu\right) u_{j,ij} + \mu u_{i,jj} - \left(\frac{\theta_0 \beta}{\rho c_\varepsilon}\right) s_{,i} = 0 \tag{26a}$$

$$\left(\frac{k}{\rho c_\varepsilon} \frac{\lambda + 2\mu}{\lambda + \frac{\theta_0 \beta^2}{\rho c_\varepsilon} + 2\mu}\right) s_{,jj} = \dot{s} \tag{26b}$$

where, for simplicity,

$$\beta = (3\lambda + 2\mu)\alpha$$

Interestingly, eqn (26b) suggests that molecular disorder, quantified by entropy density, diffuses through a CQT body. The rate at which this process progresses at any time depends upon the existing entropy density distribution and a material parameter c. This thermal diffusivity is defined by

$$c = \frac{k}{\rho c_\varepsilon} \frac{\lambda + 2\mu}{\lambda + \frac{\theta_0 \beta^2}{\rho c_\varepsilon} + 2\mu} \tag{27}$$

4.2.3 Poroelastic–Thermoelastic Analogy
Hopefully, from the presentations of the previous two sections, it has become evident that there is a strong analogy between poroelasticity and

thermoelasticity. It appears that this analogy was first discovered by Biot (1956), who earlier pioneered much of the work in poroelasticity.

Returning to eqns (12) and (21), one can see the similar roles played by the pore pressure and the temperature. However, the correspondence runs much deeper, and as Biot (1958) noted, the two phenomena are completely analogous due to the fact that both obey the same basic thermodynamic principles. Actually, the more fundamental relationships can be uncovered by comparing eqns (16) with eqns (26). In particular, this leads to the correspondence between fluid mass content (m) and entropy density (s).

Many additional relationships can be established by conducting a term-by-term investigation of eqns (16) versus eqns (26). For example, consider the coefficients of $u_{j,ij}$. In this case, the terms $\lambda + \theta_0 \beta^2 / \rho c_\varepsilon$ in CQT correspond to the undrained Lame modulus λ_u of poroelasticity. Actually this is not unexpected, since under the enforcement of isentropic deformation the relevant Lame constant, say λ_s, becomes equivalent to $\lambda + \theta_0 \beta^2 / \rho c_\varepsilon$. Thus, isentropic deformation in CQT is the counterpart of undrained deformation in poroelasticity.

Table 2 provides a detailed summary of the complete analogy. In the

TABLE 2
SUMMARY OF THE ANALOGY

Thermoelasticity	Correspondence		Poroelasticity
Entropy density	s	m	Fluid mass content per unit volume
Temperature	θ	p	Pore pressure
Heat flux	q	q	Fluid mass flux
Diffusivity	c	c	Diffusivity
$c = \dfrac{k}{\rho c_\varepsilon}\left(\dfrac{\lambda + 2\mu}{\lambda_s + 2\mu}\right)$			$c = \kappa\left(\dfrac{\lambda_u - \lambda}{\beta^2}\right)\left(\dfrac{\lambda + 2\mu}{\lambda_u + 2\mu}\right)$
Isentropic Lame constant	λ_s	λ_u	Undrained Lame constant
$\lambda_s = \lambda + \dfrac{\theta_0 \beta^2}{\rho c_\varepsilon}$			$\lambda_u = \lambda + \dfrac{B\beta K}{1 - B\beta}$
Thermoelastic constant	β	β	Poroelastic constant
$\beta = (3\lambda + 2\mu)\alpha$			$\beta = \dfrac{3}{B}\left(\dfrac{\lambda_u - \lambda}{3\lambda_u + 2\mu}\right)$
Conductivity over reference temperature	$\dfrac{k}{\theta_0}$	κ	Permeability

present work, this analogy is exploited to the fullest, so that the boundary element formulation is directly applicable to both poroelastic and thermoelastic. However, as a matter of convenience, poroelastic terminology will be used throughout the remainder of this chapter, except for several thermoelastic examples included in Section 4.6.

4.3 FUNDAMENTAL SOLUTIONS

4.3.1 Three-dimensional

A key ingredient in a boundary element formulation is the appropriate fundamental solution of the governing differential equations. In particular, for three-dimensional analysis, the infinite space, unit step point force and point source solutions of eqns (12) are required. Nowacki (1966), working within the context of coupled quasistatic thermoelasticity, appears to have been the first to develop these solutions or Green's functions. He used a decomposition of displacement into potential and solenoidal components before obtaining solutions in the Laplace transform domain. The inverse transform, then, produced the desired fundamental solutions. More recently, Cleary (1977), in an apparently independent effort, derived the same Green's functions in a much more direct manner. Cleary, working from the poroelastic vantage point, utilised physical reasoning and symmetry arguments to aid in the derivation. Subsequently, Rudnicki (1981) corrected some errors in Cleary's solution. The resulting corrected Green's functions are presented below.

Consider, first, the effect of a unit step force in the j-direction acting at the point ξ in an infinite medium. Thus, the forcing function can be written

$$f_j(\xi, t) = \delta(\xi_1)\delta(\xi_2)\delta(\xi_3)H(t)e_j \tag{28}$$

where $\delta(\xi_i)$ and $H(t)$ are standard Dirac delta and Heaviside step functions, and e_j is a unit vector. In the coupled theory, this force produces the following displacement and pore pressure at any point X:

$$u_i(X, t) = \frac{1}{16\pi r}\left[\frac{1}{\mu(1-\nu)}\right]\left[\frac{y_i y_j}{r^2}g_1(\eta) + (\delta_{ij})g_2(\eta)\right]e_j \tag{29a}$$

$$p(X, t) = \frac{1}{4\pi}\left[\frac{\beta}{\kappa(\lambda+2\mu)}\right]\left[\frac{y_j}{r}g_3(\eta)\right]e_j \tag{29b}$$

where

$$y_i = x_i - \xi_i \tag{30a}$$

$$r^2 = y_i y_i \tag{30b}$$

$$\eta = \frac{r}{(ct)^{1/2}} \tag{30c}$$

The evolution functions $g_1(\eta)$, $g_2(\eta)$ and $g_3(\eta)$ appearing in eqns (29) are

$$g_1(\eta) = 1 + c_1 [h_1(\eta) - h_2(\eta)] \tag{31a}$$

$$g_2(\eta) = (3 - 4v) - c_1 \left[h_1(\eta) - \frac{h_2(\eta)}{3} + \frac{2\eta}{3\sqrt{\pi}} e^{-\eta^2/4} \right] \tag{31b}$$

$$g_3(\eta) = \frac{h_1(\eta)}{\eta^2 t} \tag{31c}$$

in which

$$c_1 = \frac{v_u - v}{1 - v_u} \tag{32}$$

$$h_1(\eta) = \operatorname{erf}\left(\frac{\eta}{2}\right) - \frac{\eta}{\sqrt{\pi}} e^{-\eta^2/4} \tag{33a}$$

$$h_2(\eta) = \frac{6h_1(\eta)}{\eta^2} - \frac{\eta}{\sqrt{\pi}} e^{-\eta^2/4} \tag{33b}$$

with the error function defined by

$$\operatorname{erf}(z) = \frac{2}{\sqrt{\pi}} \int_0^z e^{-x^2} \, dx \tag{34}$$

The functions $g_1(\eta)$, $g_2(\eta)$ and $g_3(\eta)$ allow the fundamental solution, defined in eqns (29), to evolve naturally from undrained to drained response.

The other fundamental solution that is needed is that due to a unit step fluid mass source acting, again, at point ξ in an infinite medium. This mass source is, then,

$$\psi(\xi, t) = \delta(\xi_1)\delta(\xi_2)\delta(\xi_3)H(t) \tag{35}$$

and the corresponding displacement and pore pressure response at point

X can be written:

$$u_i(X, t) = \frac{1}{8\pi} \left[\frac{\beta}{\kappa(\lambda + 2\mu)} \right] \left[\frac{y_i}{r} g_4(\eta) \right] \tag{36a}$$

$$p(X, t) = \frac{1}{4\pi r} \left(\frac{1}{\kappa} \right) g_5(\eta) \tag{36b}$$

where

$$g_4(\eta) = \text{erfc} \left(\frac{\eta}{2} \right) + \frac{2h_1(\eta)}{\eta^2} \tag{37a}$$

$$g_5(\eta) = \text{erfc} \left(\frac{\eta}{2} \right) \tag{37b}$$

The complementary error function, erfc, is expressible as

$$\text{erfc}(z) = 1 - \text{erf}(z) \tag{38}$$

Again, the evolution functions $g_4(\eta)$ and $g_5(\eta)$ provide the transition from undrained to drained response.

4.3.2 Two-dimensional

Many physical problems can be idealised as primarily two-dimensional, within the limits of typical engineering approximation. In such cases, significant computational savings result. For that reason, a two-dimensional plane strain boundary element method will also be developed herein for the poroelastic problem. Naturally, the two-dimensional plane strain fundamental solution is required. Following the work of Rice and Cleary (1976) and Rudnicki (1987) the appropriate Green's functions are detailed below.

First, consider the effect of unit step forces acting in the j-direction at the points (ξ_1, ξ_2, x_3) where the range of x_3 is from $-\infty$ to $+\infty$. Thus, the total applied force is a line load and the desired solution can be obtained by integrating the three-dimensional Green's function (29) along the entire x_3 direction. The response at any point $(x_1, x_2, 0)$ is given by

$$u_i(X, t) = \frac{1}{8\pi} \left[\frac{1}{\mu(1 - \nu)} \right] \left[\frac{y_i y_j}{r^2} \bar{g}_1(\eta) + (\delta_{ij}) \bar{g}_2(\eta) \right] e_j \tag{39a}$$

$$p(X, t) = \frac{r}{2\pi} \left[\frac{\beta}{\kappa(\lambda + 2\mu)} \right] \left[\frac{y_j}{r} \bar{g}_3(\eta) \right] e_j \tag{39b}$$

where

$$\bar{g}_1(\eta) = 1 + c_1[1 - \bar{h}_1(\eta)] \tag{40a}$$

$$\bar{g}_2(\eta) = -(3 - 4v)\ln r + c_1[\bar{h}_2(\eta)] \tag{40b}$$

$$\bar{g}_3(\eta) = \frac{\bar{h}_1(\eta)}{4t} \tag{40c}$$

and

$$\bar{h}_1(\eta) = \frac{4}{\eta^2}(1 - e^{-\eta^2/4}) \tag{41a}$$

$$\bar{h}_2(\eta) = \frac{1}{2}\left[E_1\left(\frac{\eta^2}{4}\right) + \ln(r^2) + \bar{h}_1 \right] \tag{41b}$$

The function E_1 is the exponential integral defined by

$$E_1(z) = \int_z^\infty \frac{e^{-x}}{x}\,dx \tag{42}$$

Obviously, for this two-dimensional case, the indices i and j assume only the values 1 and 2.

Next, the continuous line fluid mass source Green's function is presented. Again, this can be obtained by integrating the three-dimensional fundamental solution expressed in eqns (36). The resulting displacement and pore pressure fields are

$$u_i(X, t) = \frac{r}{4\pi}\left[\frac{\beta}{\kappa(\lambda + 2\mu)} \right]\left[\frac{y_i}{r}\bar{g}_4(\eta) \right] \tag{43a}$$

$$p(X, t) = \frac{1}{2\pi}\left(\frac{1}{\kappa}\right)[\bar{g}_5(\eta)] \tag{43b}$$

where

$$\bar{g}_4(\eta) = \frac{\bar{h}_1(\eta)}{2} + \frac{E_1\eta^2/4}{2} \tag{44a}$$

$$\bar{g}_5(\eta) = \frac{E_1\eta^2/4}{2} \tag{44b}$$

Armed with these fundamental solutions, a boundary element formulation for coupled quasistatic poroelasticity will be developed in the following section.

4.4 BOUNDARY INTEGRAL FORMULATION

A reciprocal theorem generally provides a convenient starting point for a direct boundary element formulation. For the coupled problem at hand, it appears that Ionescu-Cazimir (1964) was the first to state explicitly an appropriate reciprocal theorem, although certainly the groundwork was laid earlier by Biot (1959). In the context of quasistatic poroelasticity, that theorem can be written, for a body of volume V and surface S with zero initial conditions, in the following time-domain form:

$$\int_s [\dot{t}_i^{(1)}*u_i^{(2)} + q^{(1)}*p^{(2)} - \dot{u}_i^{(1)}*t_i^{(2)} - p^{(1)}*q^{(2)}]\, dS$$

$$+ \int_v [\dot{f}_i^{(1)}*u_i^{(2)} + \psi^{(1)}*p^{(2)} - \dot{u}_i^{(1)}*f_i^{(2)} - p^{(1)}*\psi^{(2)}]\, dV = 0 \qquad (45)$$

The superscripts $^{(1)}$ and $^{(2)}$ denote any two independent states existing in the body defined by $[\dot{u}_i^{(1)}, \dot{t}_i^{(1)}, p^{(1)}, q^{(1)}, \dot{f}_i^{(1)}, \psi^{(1)}]$ and $[u_i^{(2)}, t_i^{(2)}, p^{(2)}, q^{(2)}, f_i^{(2)}, \psi^{(2)}]$, respectively. The symbol $*$ indicates a Riemann convolution integral, where, for example,

$$q^{(1)}*p^{(2)} = \int_0^t q^{(1)}(X, t-\tau)p^{(2)}(X, \tau)\, d\tau = \int_0^t q^{(1)}(X, \tau)p^{(2)}(X, t-\tau)\, d\tau \quad (46)$$

It should be mentioned that the form of eqn (45) above will lead to a boundary element formulation with displacement, traction, pore pressure, and flux as primary quantities. Other researchers, notably Predeleanu (1981) and Sládek and Sládek (1984b), have written equally valid reciprocal statements that, unfortunately, lead to a much less desirable set of primary variables including, for example, displacement and traction rates.

This reciprocal theorem involves any two independent states existing in the body of interest. Let one of those states, say state (2), be the, as yet, unknown solution to a given boundary value problem. Then the remaining state may be chosen arbitrarily. However, by initially selecting for state (1), the infinite region response to a unit step force in the j-direction acting at

ζ within V and beginning at time zero, the volume integrals in the reciprocal theorem can be made to vanish. That is, in eqn (45), at any point Z in V, let the applied forces and sources equal

$$f_i^{\{1\}}(Z, t) = \delta(Z - \zeta) H(t) \delta_{ij} e_j \qquad (47a)$$

$$\psi^{(1)}(Z, t) = 0 \qquad (47b)$$

and represent the response by

$$u_i^{\{1\}}(X, t) = G_{ij}(X - \zeta, t) e_j \qquad (47c)$$

$$p^{(1)}(X, t) = \dot{G}_{pj}(X - \zeta, t) e_j \qquad (47d)$$

$$t_i^{\{1\}}(X, t) = F_{ij}(X - \zeta, t) e_j \qquad (47e)$$

$$q^{(1)}(X, t) = \dot{F}_{pj}(X - \zeta, t) e_j \qquad (47f)$$

In this notation, the subscript p in eqns (47d) and (47f) does not vary, but instead takes the value 4 for three-dimensional regions, or 3 for two-dimensional problems. Obviously, eqns (47c) and (47d) are the same fundamental solutions that were presented in the previous sections, but now with a different nomenclature. For example, in the three-dimensional case, the function $G_{ij}(X - \zeta, t)$ is defined by eqn (29a), while $\dot{G}_{pj}(X - \zeta, t)$ is contained in eqn (29b). On the other hand, the functions $F_{ij}(X - \zeta, t)$ and $F_{pj}(X - \zeta, t)$ can be obtained from G_{ij} and G_{pj} through the constitutive relationships. The superposed dots in eqns (47d) and (47f) represent, as before, time derivatives which have been introduced for notational convenience. Ultimately, only the kernel functions G_{ij}, G_{pj}, F_{ij} and F_{pj} will be required in explicit form, and these are detailed for both two- and three-dimensional domains in Dargush (1987) and Dargush and Banerjee (1989a,b).

Next, for simplicity, assume that no body forces or fluid mass sources exist in the actual boundary value problem. Thus, for all Z,

$$f_i^{\{2\}}(Z, t) = 0 \qquad (48a)$$

$$\psi^{(2)}(Z, t) = 0 \qquad (48b)$$

Upon simplifying the volume integral, factoring out the common e_j term, and dropping the superscript (2), eqn (45) can be rewritten

$$\delta_{ij} u_i(\zeta, t) = \int_s [\dot{G}_{ij}(X - \zeta, t) * t_i(X, t) + \dot{G}_{pj}(X - \zeta, t) * q(X, t)$$

$$- \dot{F}_{ij}(X - \zeta, t) * u_i(X, t) - \dot{F}_{pj}(X - \zeta, t - \tau) * p(X, t)] \, dS(X) \qquad (49)$$

This is an integral equation for interior displacement that involves only boundary quantities. Therefore, volume integration has, indeed, been eliminated through the use of an infinite space Green's function.

Evidently, eqn (49) is the desired expression for the displacement vector at any interior point; however, a similar relationship is also desired for interior pore pressure. To that end, return to the reciprocal theorem (45) and select, instead, for state (1) the infinite space response to a unit pulse fluid mass source, acting at time zero and at point ξ within V. That is, let

$$f_i^{(1)}(Z, t) = 0 \tag{50a}$$

$$\psi^{(1)}(Z, t) = \delta(Z - \xi)\delta(t) \tag{50b}$$

and consequently,

$$u_i^{(1)}(X, t) = G_{ip}(X - \xi, t) \tag{50c}$$

$$p^{(1)}(X, t) = \dot{G}_{pp}(X - \xi, t) \tag{50d}$$

$$t_i^{(1)}(X, t) = F_{ip}(X - \xi, t) \tag{50e}$$

$$q^{(1)}(X, t) = \dot{F}_{pp}(X - \xi, t) \tag{50f}$$

Since this is the solution due to a unit pulse, the functions G_{ip} and $\dot{G}_{pp}$ are the time derivatives of the unit step Green's functions presented previously. In the end, it will be the kernel functions G_{ip}, G_{pp}, F_{ip} and F_{pp} that will be of primary importance, rather than their time derivatives. Again, these can be found in Dargush (1987) and Dargush and Banerjee (1989a,b).

Now, state (2) is chosen to be the actual problem, in which body forces and fluid mass sources are absent. Thus, with

$$f_i^{(2)}(Z, t) = 0 \tag{51a}$$

$$\psi^{(2)}(Z, t) = 0 \tag{51b}$$

the reciprocal theorem reduces to

$$p(\xi, t) = \int_s \left[\dot{G}_{ip}(X - \xi, t)^* t_i(X, t) + \dot{G}_{pp}(X - \xi, t)^* q(X, t) \right.$$

$$\left. - \dot{F}_{ip}(X - \xi, t)^* u_i(X, t) - \dot{F}_{pp}(X - \xi, t)^* p(X, t) \right] dS(X) \tag{52}$$

This, of course, is the desired statement for interior pore pressure in terms of boundary quantities.

Equations (49) and (52) can now be combined and rewritten in a more convenient matrix notation as

$$
\begin{Bmatrix} u_j(\xi, t) \\ p(\xi, t) \end{Bmatrix} = \int_s \left(\begin{bmatrix} \dot{G}_{ij} & \dot{G}_{ip} \\ \dot{G}_{pj} & \dot{G}_{pp} \end{bmatrix}^{\mathrm{T}} * \begin{Bmatrix} t_i(X, t) \\ q(X, t) \end{Bmatrix} - \begin{bmatrix} \dot{F}_{ij} & \dot{F}_{ip} \\ \dot{F}_{pj} & \dot{F}_{pp} \end{bmatrix}^{\mathrm{T}} * \begin{Bmatrix} u_i(X, t) \\ p(X, t) \end{Bmatrix} \right) dS(X) \quad (53)
$$

However, this can be compacted even further by generalising the displacement and traction vectors to include pore pressure and flux, respectively, as an additional component. Thus, in three dimensions, for example, let

$$
\mathbf{u}_\alpha = \{u_1 \quad u_2 \quad u_3 \quad p\}^{\mathrm{T}} \tag{54a}
$$

$$
\mathbf{t}_\alpha = \{t_1 \quad t_2 \quad t_3 \quad q\}^{\mathrm{T}} \tag{54b}
$$

where the subscript α, and subsequently β, varies from 1 to 4. Obviously, for the two-dimensional counterpart

$$
\mathbf{u}_\alpha = \{u_1 \quad u_2 \quad p\}^{\mathrm{T}} \tag{54c}
$$

$$
\mathbf{t}_\alpha = \{t_1 \quad t_2 \quad q\}^{\mathrm{T}} \tag{54d}
$$

Then eqn (53) becomes simply

$$
\mathbf{u}_\alpha(\xi, t) = \int_s [\dot{\mathbf{G}}_{\beta\alpha} * \mathbf{t}_\beta(X, t) - \dot{\mathbf{F}}_{\beta\alpha} * \mathbf{u}_\beta(X, t)] \, dS(X) \tag{55}
$$

Equation (55) can be viewed as a generalised Somigliana's identity for quasistatic poroelasticity, and, as such, is an exact statement for the interior displacements and pore pressure at any point ξ within V at any time t. However, to determine those interior quantities, the entire history of boundary values of $\mathbf{u}_\alpha$ and $\mathbf{t}_\alpha$ must be known. Unfortunately, in a well-posed boundary value problem only half of that information is given at each instant of time. In order to obtain the missing information, and in essence to solve the problem, the point ξ must be moved to the boundary.

Due to the singular nature of the kernel functions, the process of writing eqn (55) for a point on the boundary is not straightforward, and considerable care must be exercised in evaluating the surface integrals. The order of the singularities in the various components of $\mathbf{G}_{\beta\alpha}$ and $\mathbf{F}_{\beta\alpha}$ is defined in Table 3. Symbolically, the boundary integral equation can be written simply as

$$
\mathbf{c}_{\beta\alpha}(\xi)\mathbf{u}_\beta(\xi, t) = \int_s [\dot{\mathbf{G}}_{\beta\alpha} * \mathbf{t}_\beta(X, t) - \dot{\mathbf{F}}_{\beta\alpha} * \mathbf{u}_\beta(X, t)] \, dS(X) \tag{56}
$$

where $\mathbf{c}_{\beta\alpha}(\xi)$ is a matrix of constants, dependent only upon the local

TABLE 3
SINGULARITIES IN THE POROELASTIC KERNELS

Component	Three-dimensional	Two-dimensional
G_{ij}	$1/r$	$\ln r$
G_{ip}	non-singular	non-singular
G_{pj}	non-singular	non-singular
G_{pp}	$1/r$	$\ln r$
F_{ij}	$1/r^2$	$1/r$
F_{ip}	non-singular	non-singular
F_{pj}	$1/r$	$\ln r$
F_{pp}	$1/r^2$	$1/r$

geometry of the boundary at ζ. For ζ on a smooth surface, $c_{\beta\alpha}$ reduces to $\delta_{\beta\alpha}/2$.

In principle, at each instant of time progressing from time zero, this equation can be written at every point on the boundary. The collection of the resulting equations could then be solved simultaneously, producing exact values for all the unknown boundary quantities. In reality, of course, discretisation is needed to limit this process to a finite number of equations and unknowns. Techniques useful for the discretisation of eqn (56) are the subject of the following section.

4.5 NUMERICAL IMPLEMENTATION

The boundary integral equation (56), developed in the last section, is an exact statement. No approximations have been introduced other than those used to formulate the boundary value problem described by eqns (12)–(15). However, in order to apply eqn (56) to the solution of practical engineering problems, approximations are required in both time and space.

In this section, highlights of a general-purpose, state-of-the-art numerical implementation are presented. This entire implementation has been accomplished within the GP-BEST (General Purpose Boundary Element Solution Technique) analysis system, which, in addition to poroelasticity, contains a wide spectrum of capabilities in linear and non-linear solid and fluid mechanics. Some of these have already been described in Chapters 1 to 3 of this volume. As a result, many of the features and techniques used in this implementation were developed previously for elastostatics and elastodynamics (e.g. Banerjee et al., 1985, 1986, 1988; Banerjee and Ahmad, 1985; Ahmad and Banerjee, 1988; and Chapters 1–3), but are here adapted for poroelastic analysis. Details of this implementation in the context of poroelastic analyses can be found in Dargush (1987) and Dargush and Banerjee (1989*a, b, c*). In this subsection, a number of these advanced features of the general purpose implementation are briefly mentioned.

Perhaps the most significant of these items is the capability to analyse substructured problems. This not only extends the analysis to bodies composed of several different materials, but also often provides computational efficiencies. During the integration process, the individual regions remain separate entities. The generic modelling regions (or GMR's) are brought together for the first time at the assembly stage, where compatibility relations are enforced on the common boundary between adjacent regions. This multi-GMR assembly process produces block-banded system matrices that can be solved in an efficient manner.

As another convenience, a high degree of flexibility is provided for the specification of boundary conditions. In general, time-dependent values can be defined in either global or local coordinates, with or without sliding. Not only can generalised displacements and tractions be specified, but also spring and smearing contact resistance relations, in the form of eqns (13c) and (14c), respectively, are available. Furthermore, a comprehensive symmetry capability is included with provisions for both planar and cyclic symmetry.

A final item worthy of mention is the availability of enclosing elements (Ahmad and Banerjee, 1988; Henry and Banerjee, 1988) in both two and three dimensions. This is a particular benefit in many soil problems involving semi-infinite regions. In such cases, the number of system equations can be considerably reduced and yet the strongly singular diagonal blocks of the F matrix can be accurately determined.

These advanced features, while not required for a boundary element implementation, greatly extend the range of applicability of the present formulation. Hopefully, this will become evident in the next section, where many of the above items are utilised in the solution of a variety of illustrative examples. Included are applications in both poroelasticity and thermoelasticity.

4.6 APPLICATIONS

4.6.1 Consolidation under a Strip Load

A number of analytical, as well as finite element, solutions are available for the consolidation of a single poroelastic stratum beneath a strip load. The geometry for this problem is completely defined in the x–y plane by the depth of the layer H, and the total breadth of the strip load $2a$. The lower boundary remains impervious, while free drainage is permitted along the entire upper surface. For convenience, non-dimensional forms of the material parameters are utilised. In particular, let $E = 1.0$, $v = 0.0$, $\kappa = 1.0$, $v_u = 0.5$, and $B = 1.0$. Notice, for this set of properties, that the diffusivity is unity. The forthcoming presentation of results is simplified further by also assuming unit values for both the half-width a and the applied traction.

Analytical solutions for this general problem were developed by Gibson et al. (1970), although previous work for the special case of infinite H is contained in Schiffman et al. (1969). The finite element results included in the following graphs are those of Yousif (1984).

Meanwhile, the boundary element mesh for the particular case of $H/a = 5$ is depicted in Fig. 1. It contains 12 elements, 25 boundary nodes, and nine interior points. Symmetry is imposed along $x = 0$. As in the finite element analysis cited above, the region is truncated at a horizontal dimension of $L/a = 10$. While theoretically the vertical elements along $x = 10$ are unnecessary, closure of the region permits accurate determination of the strongly singular diagonal blocks of the **F** matrix via rigid body considerations. A recently discovered alternative is the enclosing element concept, which was developed by Ahmad and Banerjee (1988) for elastodynamics.

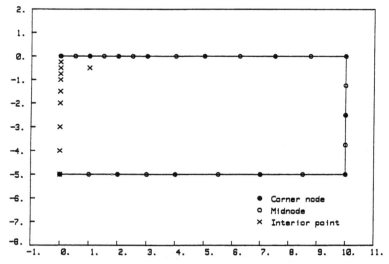

FIG. 1   Consolidation under a strip load, boundary element model.

With this new technique, the enclosing elements are used only in the calculations of the diagonal blocks of **F**, and do not produce additional boundary equations.

First, results are presented for $H/a \rightarrow \infty$. Figure 2 contains the pore

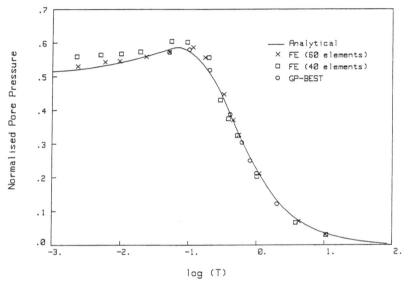

FIG. 2   Consolidation under a strip load, results showing normalised pore pressure vs log T.

pressure time history for a point, on the centreline, a unit depth below the surface. The well-known Mandel–Cryer effect, of increasing pore pressure during the early stages of the process, is evident in the graph. Actually, only the final portion of the rising pore pressure response was captured in the boundary element analysis, since a relatively large time step was selected. Even with this large step, however, the analytical solution is accurately reproduced. In Fig. 3, the pore pressure distribution with depth is shown for $T=0.1$. Obviously, the boundary element solutions are comparable to those obtained by Yousif (1984) via finite elements.

Next, Fig. 4 displays the degree of consolidation versus time for various H/a ratios. Again, in all cases, excellent agreement with the analytical solution is obtained.

Finally, the effects of partial saturation are explored. For soils, this is accomplished by maintaining the effective stress parameter β at unity, while

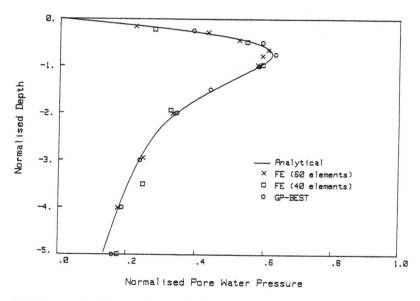

FIG. 3 Consolidation under a strip load, results showing normalised depth vs normalised pore water pressure.

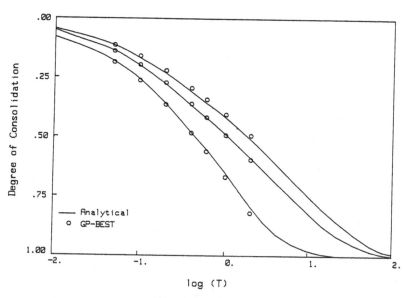

FIG. 4 Consolidation under a strip load, results showing degree of consolidation vs log T.

reducing Skempton's coefficient B to various levels. Consequently, the undrained Poisson's ratio will assume values less than 0.5. Physically, as the degree of saturation is reduced, v_u and B decrease, resulting in a larger immediate settlement and a slower consolidation process. This behaviour is evident in the boundary element results presented in Figs 5 and 6 for the particular case of $H/a = 10$. Analytical solutions are unavailable for this partially saturated problem.

4.6.2 Consolidation at the Koyna Dam

The next example involves the simulation of a consolidation process beneath the Koyna Dam and reservoir in India. The objective of the analysis is to determine the changes in excess pore pressures and effective stresses at various subsurface locations due to impoundment and seasonal variations in reservoir level. All geometric, load and material property definitions were provided by Roeloffs (1985).

As a first approximation, the reservoir is considered as a uniformly loaded 40 km by 5 km rectangle resting on a two-layer half-space. Specifically, the subsurface geology is idealised by a 1 km thick zone of basalt on top of a semi-infinite granite region. The material properties that

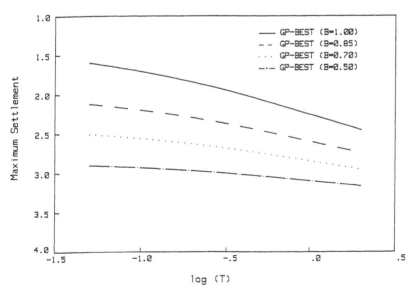

FIG. 5   Consolidation under a strip load, results ($H/a = 10$) for maximum settlement.

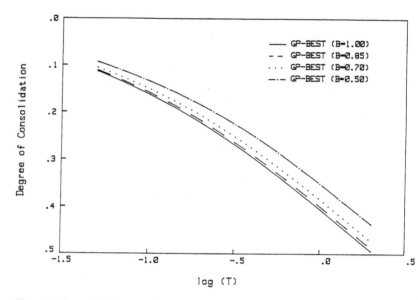

FIG. 6 Consolidation under a strip load, results ($H/a = 10$) for degree of consolidation.

were adopted for each layer are detailed in Table 4. The reservoir level, and consequently, the load history, is a function of time, involving the initial filling stage plus typical seasonal variations. Table 5 contains the assumed reservoir levels, h, during the first three years after impoundment. This information is used to determine the boundary conditions on the outer surface. Beneath the reservoir, let

$$t_y = -p = \rho g h$$
$$t_x = t_z = 0$$

TABLE 4

KOYNA DAM SUBSURFACE MATERIAL PROPERTIES

	E (GPa)	v	κ (km/year)	v_u	B
Basalt	75·0	0·25	$2·0 \times 10^{-3}$	0·3	0·8
Granite	75·0	0·25	$2·0 \times 10^{-5}$	0·3	0·8

TABLE 5
KOYNA DAM RESERVOIR LEVELS

t (months)	h (m)	t (months)	h (m)
0	0·0	19	8·9
2	2·3	24	6·5
7	5·0	26	13·3
12	4·8	31	10·4
14	10·3	36	7·2

while elsewhere

$$t_x = t_y = t_z = p = 0$$

In the above, ρg is the weight density of water.

Due to the elongated shape of the reservoir, a two-dimensional plane strain analysis, taken in the plane normal to the long axis, is in order. The two-region boundary element mesh is detailed in Fig. 7. The basalt layer is modelled with 17 elements, while only 12 elements are needed to describe the response of the granite. An ample supply of interior nodes is also included to monitor excess pore pressures and effective stresses throughout the area of interest. Results for points A, B, C and D are presented in Fig. 8.

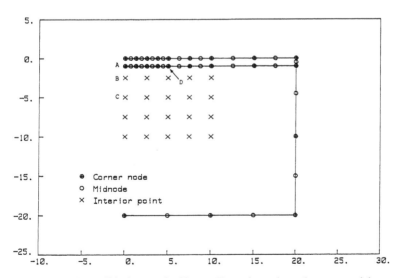

FIG. 7   Consolidation at the Koyna Dam, boundary element model.

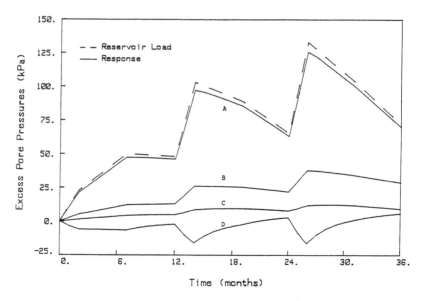

FIG. 8   Consolidation at the Koyna Dam, plane strain results showing development over time of excess pore pressures.

In that diagram, the excess pore pressures at those four locations are traced versus time. Notice that the intensities drop dramatically with depth and that, in some instances, load reversals produce negative pore pressures in the underlying strata. It has been hypothesised by some that consolidation, as considered here, could induce seismic events. However, in all cases, the magnitudes of the pressures and stresses due to this consolidation process are very small, particularly when compared to typical tectonic stresses.

4.6.3 Consolidation under Rectangular Loads
This last poroelastic application involves the fully three-dimensional problem of the consolidation of a half-space under rectangular loads of various dimensions. The three specific cases considered involve length (b) to width (a) ratios of 1·0, 2·0 and 8·0 for the loaded area. A boundary element model corresponding to the latter case is shown in Fig. 9. Notice that with the present formulation, discretisation is required on only a small portion of the half-space. Additionally, quadrantal symmetry is utilised to reduce computational effort.
Figure 10 displays the results for the degree of consolidation at the centre

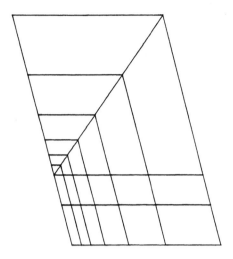

FIG. 9   Consolidation under a load ($b/a=8$), boundary element model.

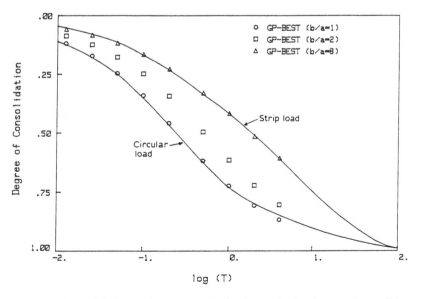

FIG. 10   Consolidation under rectangular loads, results for degree of consolidation.

of the loaded area versus non-dimensional time. For comparative purposes, the analytical curves of Gibson *et al.* (1970) for consolidation under strip and circular loads are also shown. It is evident, from the graph, that the behaviour under a rectangular load with an aspect ratio of 8 is nearly identical to the response due to a strip load. At the other end of the spectrum, the results for a square load closely follow the analytical solution for a circular load, with slight deviation present at large times. As expected, the boundary element results for $b/a = 2$ fall between the two limiting cases.

4.6.4 Cooling of a Steel Sphere

Shifting now to the thermoelastic arena, consider a solid sphere of radius 1·5 in, initially at a uniform temperature of 200° F, cooled in 100° F air by convection. The thermomechanical properties of the steel are assumed to be as follows:

$$E = 30·0 \times 10^6 \text{ psi} \qquad \rho c_\varepsilon = 283 \text{ in lb/in}^3 \,°\text{F}$$
$$\nu = 0·30 \qquad\qquad k = 5·8 \text{ in lb/s in } °\text{F}$$
$$\alpha = 6·0 \times 10^{-6} \text{ in/in} °\text{F}$$

Three levels of convection are examined: very slow cooling ($h = 1·25 \text{ in lb/s in}^2 °\text{F}$), moderate cooling ($h = 20·0 \text{ in lb/s in}^2 °\text{F}$) and rapid cooling ($h \to \infty$). Analytical solutions for the temperature distribution in the sphere can be found in Carslaw and Jaeger (1947), while the corresponding stress fields for the rapid cooling case are presented in Nowacki (1986).

The GPBEST model for this problem consists of two eight-noded and two six-noded surface elements on an octant of the sphere. As a result, a total of 17 source points are utilised, along with the octahedral symmetry option. A time step of 1·25 s is selected for all analyses.

The resulting temperatures versus time for the point at the centre of the sphere are shown in Fig. 11. It is apparent that an excellent correlation is obtained for slow, as well as rapid, cooling. Meanwhile, Fig. 12 displays the radial stresses at that point. The correlation is again very good throughout the time history. Notice that, as expected, with slower surface cooling, the peak stress magnitude decreases while the time to reach steady-state (i.e. zero stress) increases. Finally, in Fig. 13, the tangential surface stresses are plotted. Even in the severe, rapid cooling case, the GPBEST results are quite accurate. In fact, this highlights one of the primary advantages of the method for transient thermoelastic problems. The steep thermal gradients that are present at the initial instant can be captured with a high degree of

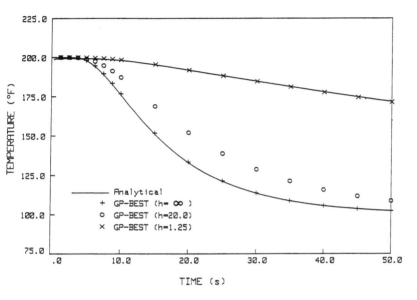

FIG. 11   Cooling of a steel sphere, GPBEST results for temperature.

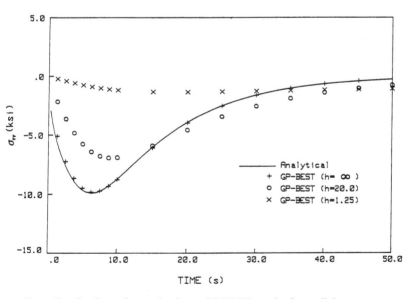

FIG. 12   Cooling of a steel sphere, GPBEST results for radial stresses.

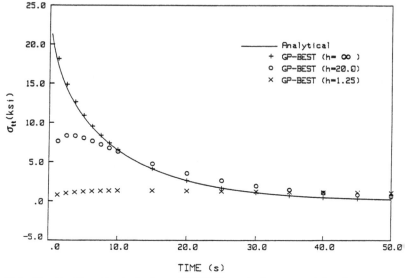

FIG. 13   Cooling of a steel sphere, GPBEST results for tangential surface stresses.

precision. Consequently, the thermal surface stresses can also be calculated accurately.

4.6.5 Turbine Blade

For the final application, the plane strain response of an internally cooled turbine blade is examined under start-up thermal transients. The boundary element model of the blade is illustrated in Fig. 14. In this problem, the two-GMR approach is chosen solely to enhance computational efficiency. This is accomplished by reducing the aspect ratio of individual GMR's and by creating a block-banded system matrix. The leading (left-hand) GMR consists of 26 quadratic elements, while 24 elements are used to model the trailing (right-hand) region.

The blade is manufactured of stainless steel with the following thermo-mechanical properties:

$$E = 29 \cdot 0 \times 10^6 \, \text{psi} \qquad \rho c_\varepsilon = 368 \, \text{in lb/in}^3 \, ^\circ\text{F}$$
$$v = 0 \cdot 30 \qquad k = 1 \cdot 65 \, \text{in lb/s in} \, ^\circ\text{F}$$
$$\alpha = 9 \cdot 6 \times 10^{-6} \, \text{in/in} \, ^\circ\text{F}$$

During operation a hot gas flows outside the blade, while a relatively cool gas passes through the internal holes. The gas temperature transients are

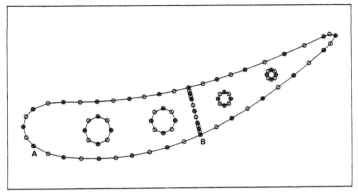

FIG. 14   Turbine blade, boundary element model.

plotted in Fig. 15 for a typical start-up. Convection film coefficients are specified as follows:

Outer surface at leading edge $h = 50 \text{ in lb/s in}^2\,°\text{F}$
Remainder of outer surface $h = 20 \text{ in lb/s in}^2\,°\text{F}$
Inner cooling hole surfaces $h = 10 \text{ in lb/s in}^2\,°\text{F}$

A time step of 0·2 is employed for the GPBEST analysis.

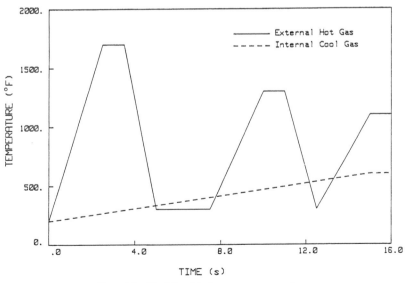

FIG. 15   Turbine blade, start-up transient.

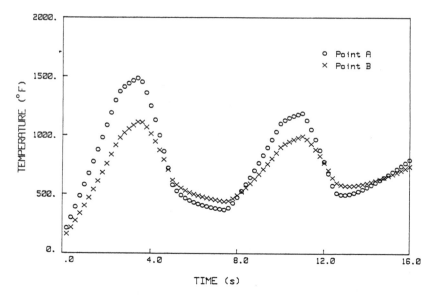

FIG. 16   Turbine blade, GPBEST results for temperature.

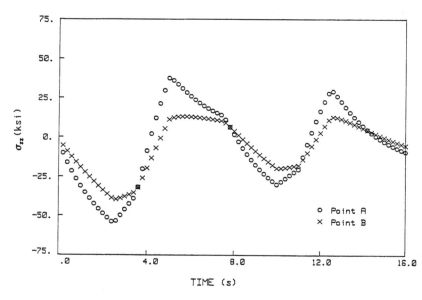

FIG. 17   Turbine blade, GPBEST results for surface stresses.

The responses at two points, A on the leading edge and B at midspan, are displayed in Figs 16 and 17. Notice that temperatures and stresses are consistently higher on the leading edge, reaching peak values of approximately $1500°F$ and -60 ksi, respectively. Also, as is evident from Fig. 17, significant stress reversals occur during this start-up. As a next step, these results from GPBEST could be used as input for a fatigue analysis to assess the durability of the design. In that regard, it should be emphasised that the stresses presented for points A and B are surface stresses, calculated by satisfying the constitutive laws, strain–displacement and equilibrium directly at the boundary point. As was mentioned in the previous example, this can be expected to produce much more accurate results than the standard practice utilised in finite element approaches of extrapolating interior Gauss point stress values to the boundary.

4.7 CONCLUSIONS

A boundary element method has been developed for application to two- and three-dimensional problems of poroelasticity and thermoelasticity. The method is formulated directly in the time domain and requires only surface discretisation. Thus, the dimensionality of the problem is reduced by one. This is particularly attractive for poroelasticity, since most practical problems involve large or semi-infinite domains.

The general purpose, state-of-the-art boundary element implementation (GPBEST) insures efficient integration, assembly and solution, while providing capability for the modelling of complicated multi-region geometry and the specification of arbitrary time-varying boundary conditions. Additionally, facilities are provided for planar symmetry and the calculation of interior and boundary pore pressures and effective stresses. A number of detailed numerical examples validate this new formulation and highlight the versatility of the implementation.

REFERENCES

AHMAD, S. & BANERJEE, P.K. (1988). Transient elastodynamic analysis of three-dimensional problems by BEM, *Int. J. Numer. Methods Eng.* **26**(8), 1560–80.
ARAMAKI, G. (1986). Boundary elements for thin layers with high permeability in Biot's consolidation analysis, *Appl. Math. Model.*, **10**, 82–6.

ARAMAKI, G. & YASUHARA, K. (1985). Application of the boundary element method for axisymmetric Biot's consolidation, *Eng. Anal.*, **2**, 184–91.

BANERJEE, P.K. & AHMAD, S. (1985). Advanced three-dimensional dynamic analysis of 3-D problems by boundary element methods, *Proc. ASME Conf. on Advanced Topics in Boundary Element Analysis*, Eds. T.A. Cruse, A.B. Pifko and H. Armen, AMD Vol. 72, ASME, New York, 65–81.

BANERJEE, P.K. & BUTTERFIELD, R. (1981). *Boundary Element Methods in Engineering Science*, McGraw-Hill, London. Also US Edition, New York, 1983; Russian Edition, Moscow, 1984.

BANERJEE, P.K. & BUTTERFIELD, R. (1982). Transient flow through porous elastic media, Chapter 2 in: *Developments in Boundary Element Methods—2*, Eds P.K. Banerjee and R.P. Shaw, Applied Science Publishers, London, 19–35.

BANERJEE, P.K. & DARGUSH, G.F. (1988). Progress in BEM applications in geomechanics via examples, *6th Int. Conf. on Numerical Methods in Geomechanics*, April 1988. Balkema, Innsbruck, Austria.

BANERJEE, P.K., WILSON, R.B. & MILLER, N.M. (1985). Development of a large BEM system for three-dimensional inelastic analysis, *Proc. ASME Conf. on Advanced Topics in Boundary Element Analysis*, Eds T.A. Cruse, A.B. Pifko and H. Armen, AMD Vol. 72, ASME, New York, 1–20.

BANERJEE, P.K., AHMAD, S. & MANOLIS, G.D. (1986). Transient elastodynamic analysis of three-dimensional problems by boundary element method, *Earthquake Eng. Structural Dyn.*, **14**, 933–49.

BANERJEE, P.K., WILSON, R.B. & MILLER, N.M. (1988). Advanced elastic and inelastic three-dimensional analysis of gas turbine engine structures by BEM, *Int. J. Numer. Methods Eng.* **26**, 393–411.

BIOT, M.A. (1956). Thermoelasticity and irreversible thermodynamics, *J. Appl. Phys.*, **27**, 240–53.

BIOT, M.A. (1958). Linear thermodynamics and the mechanics of solids, *Proc. 3rd US National Congress of Applied Mechanics*, Brown University, Providence, RI, June 1958.

BIOT, M.A. (1959). New thermomechanical reciprocity relations with application to thermal stress analysis, *J. Aero/Space Sci.* **26**(7), 401–8.

BOLEY, B.A. & WEINER, J.H. (1960). *Theory of Thermal Stresses*, John Wiley, New York.

CARSLAW, H.S. & JAEGER, J.C. (1947). *Conduction of Heat in Solids*, Clarendon Press, Oxford, UK.

CHAUDOUET, A. (1987). Three-dimensional transient thermoelastic analysis by the BIE method, *Int. J. Numer. Methods Eng.*, **24**, 25–45.

CHENG, A. H.-D. & LIGGETT, J.A. (1984). Boundary integral equation method for linear porous-elasticity with applications to soil consolidation, *Int. J. Numer. Methods Eng.*, **20**, 255–78.

CLEARY, M.P. (1977). Fundamental solutions for a fluid-saturated porous solid, *Int. J. Solids Struct.* **13**, 785–806.

CRUSE, T.A., SNOW, D.W. & WILSON, R.B. (1977). Numerical solutions in axisymmetric elasticity, *Comput. Strut.*, **7**, 445–51.

DARGUSH, G.F. (1987). Boundary element methods for the analogous problems of thermomechanics and consolidation, Ph.D. Thesis, State University of New York at Buffalo.

DARGUSH, G.F. & BANERJEE, P.K. (1989a). A time domain boundary element method for poroelasticity, *Int. J. Numer. Methods Eng.* (in press).
DARGUSH, G.F. & BANERJEE, P.K. (1989b). Development of a boundary element method for time-dependent planar thermoelasticity, *Int. J. Solids Struct.* (in press).
DARGUSH, G.F. & BANERJEE, P.K. (1989c). Boundary element methods in three-dimensional thermoelasticity, *Int. J. Solids Struct.* (in press).
DAY, W.A. (1982). Further justification of the uncoupled and quasistatic approximations in thermoelasticity, *Arch. Ration. Mech. Anal.*, 79(1), 387–96.
GIBSON, R.E., SCHIFFMAN, R.L. & PU, S.L. (1970). Plane strain and axially symmetric consolidation of a clay layer on a smooth impervious base, *Qtrly J. Mech. Appl. Math.*, 23, 505–20.
HENRY, D.P. & BANERJEE, P.K, (1988). A variable stiffness type boundary element formulation for axisymmetric elastoplastic media. *Int. J. Numer. Methods Eng.* 26, 1005–27.
IONESCU-CAZIMIR, V. (1964). Problem of linear coupled thermoelasticity. Theorems on reciprocity for the dynamic problem of coupled thermoelasticity. I, *Bull. Acad. Pol. Sci., Sér. Sci. Tech.*, 12(9), 473–88.
KUROKI, T., ITO, T. & ONISHI, K. (1982). Boundary element method in Biot's linear consolidation, *Appl. Math. Model.*, 6, 105–10.
MASINDA, J. (1984). Application of the boundary element method to 3D problems of non-stationary thermoelasticity, *Eng. Anal.*, 1, 66–9.
NISHIMURA, N. (1985). A BIE formulation for consolidation problems, in *Boundary Elements VII, Proc. 7th Int. Conf.*, Eds C.A. Brebbia and G. Maier, Springer–Verlag, Berlin.
NISHIMURA, N., UMEDA, A. & KOBAYASHI, S. (1986). Analysis of consolidation by boundary integral equation method, in *Boundary Elements, Proc. Int. Conf.*, Beijing, Ed. DuQuinghua, Pergamon, Oxford, 119–26.
NOWACKI, W. (1966). Green's functions for a thermoelastic medium (Quasistatic problem), *Bull. Inst. Polit. Jasi, Sierie noua*, 12(3–4), 83–92.
NOWACKI, W. (1986). *Thermoelasticity*, Pergamon, Warsaw.
PREDELEANU, M. (1981). Boundary integral method for porous media, in *Boundary Methods*, Ed. C.A. Brebbia, Springer–Verlag, Berlin, 325–34.
RICE, J.R. & CLEARY, M.P. (1976). Some basic stress diffusion solutions for fluid-saturated elastic porous media with compressible constituents, *Rev. Geophys. Space Phys.*, 14(2), 227–41.
RIZZO, F.J. & SHIPPY, D.J. (1977). An advanced boundary integral equation method for three-dimensional thermoelasticity, *Int. J. Numer. Methods Engr.*, 11, 1753–68.
ROELOFFS, E. (1985). Private communication.
RUDNICKI, J.W. (1981). On 'Fundamental solutions for a fluid-saturated porous solid', by M.P. Cleary, *Int. J. Solids Struct.*, 17, 855–7.
RUDNICKI, J.W. (1987). Fluid mass sources and point forces in linear elastic diffusive solids, *Mech. Mater.*, 5, 383–93.
SCHIFFMAN, R.L., CHEN, A.T.-F. & JORDAN, J.C. (1969). An analysis of consolidation theories, *J. Soil Mech. Foundations Div., Am. Soc. Civ. Eng.*, 95, SM1, 285–312.
SHARP, S. & CROUCH, S.L. (1986). Boundary integral methods for thermoelasticity problems, *J. Appl. Mech.*, 53, 298–302.

SLÁDEK, V. & SLÁDEK, J. (1983). Boundary integral equation method in thermo-elasticity, Part I: General analysis, *Appl. Math. Model.*, 7, 241–53.

SLÁDEK, V. and SLÁDEK, J. (1984a). Boundary integral equation method in thermoelasticity. Part III: Uncoupled thermoelasticity, *Appl. Math. Model.*, 8, 413–18.

SLÁDEK, V. and SLÁDEK, J. (1984b). Boundary integral equation method in two-dimensional thermoelasticity. *Eng. Anal.*, 1, 135–48.

TANAKA, M. & TANAKA, K. (1981) Boundary element approach to dynamic problems in coupled thermoelasticity—1. Integral equation formulation, *Solid Mech. Arch.*, 6, 467–91.

TANAKA, M., TOGOH, H. & KIKUTA, M. (1984). Boundary element method applied to 2-D thermoelastic problems in steady and non-steady states, *Eng. Anal.*, 1, 13–19.

YOUSIF, N.B. (1984). Finite element analysis of some time independent construction problems in geotechnical engineering, Ph.D. Thesis, State University of New York at Buffalo.

Chapter 5

ADVANCED SUBSTRUCTURED ANALYSIS OF ELASTODYNAMIC WAVE PROPAGATION PROBLEMS

N. NISHIMURA, S. KOBAYASHI

Department of Civil Engineering, Kyoto University, Kyoto, Japan

and

M. KITAHARA

Department of Ocean Engineering, Tokai University, Shimizu, Japan

SUMMARY

Various boundary integral equation (BIE) formulations in elastodynamics and viscoelastodynamics are presented. A domain-typed boundary integral equation method (BIEM) for inhomogeneous anisotropic materials and a regularised time-domain double layer potential method for crack problems are described. Combined use of these methods with finite element methods (FEM) is discussed.

5.1 INTRODUCTION

Substructured analysis refers to those methods of numerical analysis in which one subdivides the domain under consideration (structure, soil, etc.) into subdomains called substructures, uses proper numerical methods of analysis for each substructure, and then connects these component analyses together to investigate the behaviour of the whole system. There are several reasons why this method is useful. For example, this method is considered to increase the numerical accuracy when one analyses a slender structure by the boundary integral equation method (BIEM) even when the structure is mechanically homogeneous (Lachat and Watson, 1976). In this chapter, however, we are more interested in the use of the substructure method in the analysis of a system composed of substructures having a variety of mechanical and geometrical properties.

In earthquake engineering, for example, one often has to carry out dynamic interaction analysis of soil and structure. Obviously soil and structures have totally different geometrical scales. Indeed, the size of a structure is usually somewhere between several metres and several kilometres, in which length the material property may vary considerably. On the other hand, soil may often be regarded essentially as semi-infinite. In addition, it would be important to know the deformation of each member of the structure, while such detail may not be required for soil. It would therefore be ineffective to carry out an analysis for the whole system by using one single numerical method of analysis. Actually, we shall find in the sequel, that the combined use of the finite element method (FEM) for the structure and BIEM for the soil is effective for such analysis.

There are two types of investigations in the BIEM community directed towards the improvement of the substructured analysis. One devises techniques to combine BIEM with other methods, such as FEM, FDM, a different type of BIEM, etc., while the other attempts to increase the capability of BIEM as component methods of analysis, i.e. the numerical methods for substructures to be combined with other methods. The topics we are going to discuss in this chapter belong to one, or both, of these types. Specifically, we shall start this chapter with mathematical statements of the problem. We shall pay attention to engineering problems where elasto-dynamic and/or viscoelastodynamic modelling is considered reasonable. We then discuss BIE formulations for elastodynamics and viscoelasto-dynamics. In particular, we investigate domain-typed BIEM and a regu-larised double layer potential method in detail. The former is useful when the substructure is inhomogeneous, while the latter is effective when the substructure contains cracks. We then proceed to discuss the numerical methods of combining the BIEM and FEM used here. This chapter concludes with several numerical examples.

5.2 STATEMENT OF THE PROBLEMS

5.2.1 Elastodynamic problems (Eringen and Suhubi, 1975; Kobayashi and Nishimura, 1982a)

Let D be a domain in R^N ($N = 2$ or 3). Also, let the time interval T be defined by $T := (0, \infty)$.

The objective of the linear elastodynamics is to find a solution to the Navier–Cauchy equation of motion

$$\nabla \cdot (\mathbf{C}(\mathbf{x}): \nabla \mathbf{u}(\mathbf{x}, t)) + \rho(\mathbf{x})\mathbf{b}(\mathbf{x}, t) = \rho(\mathbf{x})\ddot{\mathbf{u}}(\mathbf{x}, t) \tag{1}$$

in $D \times T$, which satisfies the initial conditions

$$\mathbf{u}(\mathbf{x}, 0) = \mathbf{u}_0(\mathbf{x}) \quad \dot{\mathbf{u}}(\mathbf{x}, 0) = \dot{\mathbf{u}}_0(\mathbf{x}) \tag{2}$$

and boundary conditions

$$\mathbf{u}(\mathbf{x}, t) = \bar{\mathbf{u}}(\mathbf{x}, t) \quad (\mathbf{x}, t) \in \partial D_1 \cdot T$$
$$\mathbf{t}(\mathbf{x}, t) = \mathbf{n} \cdot \mathbf{C}: \nabla \mathbf{u}(\mathbf{x}, t) = \bar{\mathbf{t}}(\mathbf{x}, t) \tag{3}$$
$$(\mathbf{x}, t) \in \partial D_2 \times T \quad \partial D = \partial D_1 \cup \partial D_2$$

where we have used the following notation: $\mathbf{u}$ = displacement, $\mathbf{t}$ = surface traction, $\mathbf{b}$ = body force, ρ = density, $\mathbf{C}$ = elasticity tensor, ∂D = boundary of D, ∂D_1, ∂D_2 = parts of ∂D, where displacement and traction, respectively, are given, ∇ = gradient, $\mathbf{n}$ = unit outward normal vector to ∂D, $\cdot$ = time derivative, $\mathbf{u}_0$ = initial displacement, $\dot{\mathbf{u}}_0$ = initial velocity, $\bar{\mathbf{u}}$ = prescribed boundary displacement, $\bar{\mathbf{t}}$ = prescribed surface traction, $(\mathbf{C}: \nabla \mathbf{u})_{ij} = C_{ijkl}u_{k,l}$.

For an isotropic and homogeneous body, the elasticity tensor reduces to

$$C_{ijkl} = \lambda \delta_{ij}\delta_{kl} + \mu(\delta_{ik}\delta_{jl} + \delta_{il}\delta_{jk}) \tag{4}$$

Hence we have

$$\mu \Delta \mathbf{u} + (\lambda + \mu)\nabla \nabla \cdot \mathbf{u} + \rho \mathbf{b} = \rho \ddot{\mathbf{u}} \tag{5}$$

and

$$\mathbf{t} = \overset{n}{\mathbf{T}}\mathbf{u} = \lambda \mathbf{n}(\nabla \cdot \mathbf{u})\mathbf{1} + \mu \mathbf{n} \cdot (\nabla \mathbf{u} + (\nabla \mathbf{u})^{\mathrm{T}}) \tag{6}$$

where λ and μ are Lame's constants, and $\overset{n}{\mathbf{T}}$ is the operator implied in eqn (6), respectively.

5.2.2 The Reduced Elastodynamic Problems
Use of the Fourier transform with respect to time defined as

$$\hat{f}(\mathbf{x}; \omega) = \int\limits_{-\infty}^{\infty} f(\mathbf{x}, t)e^{i\omega t}\, \mathrm{d}t \tag{7}$$

converts the original elastodynamic problem into the reduced elastodynamic problem expressed as follows:

$$\nabla \cdot (\mathbf{C}: \nabla \hat{\mathbf{u}}(\mathbf{x}; \omega)) + \rho \omega^2 \hat{\mathbf{u}}(\mathbf{x}; \omega) = \mathbf{0}, \quad \mathbf{x} \in D \tag{8}$$

$$\hat{\mathbf{u}}(\mathbf{x}; \omega) = \hat{\bar{\mathbf{u}}}(\mathbf{x}; \omega), \quad \mathbf{x} \in \partial D_1 \tag{9a}$$

$$\overset{n}{\mathbf{T}} \hat{\mathbf{u}}(\mathbf{x}; \omega) = \hat{\bar{\mathbf{t}}}(\mathbf{x}; \omega), \quad \mathbf{x} \in \partial D_2 \tag{9b}$$

where ω is the parameter of the transform, and the body force is disregarded. The reduced problem is now elliptic. Therefore the solution techniques developed for elastostatics are applicable. The solutions to the original problem are obtained as the Fourier inverse transform of the solutions of the reduced problem.

In exterior boundary value problems, the solution must satisfy radiation conditions, which imply that the scattered field consists of outgoing waves (Kupradze et al., 1979).

In the sequel we shall drop ^ in reduced elastodynamic problems in order to simplify the notation.

5.2.3 Viscoelastodynamic Problems (Christensen, 1971; Kobayashi et al., 1986)

For a linear viscoelastic body, whose stress ($\boldsymbol{\sigma}$)–strain ($\boldsymbol{\varepsilon}$) relation is given by

$$\boldsymbol{\sigma}(\mathbf{x}, t) = \int_{-\infty}^{t} \mathbf{G}(t - \tau) : \frac{d\boldsymbol{\varepsilon}}{d\tau}(\mathbf{x}, \tau) \, d\tau \tag{10}$$

the reduced equation of motion takes the form of eqn (8) with $\mathbf{C}$ replaced by the complex modulus $\hat{\mathbf{G}}$ defined by

$$\hat{\mathbf{G}}(\omega) = \int_{-\infty}^{\infty} e^{i\omega t} \mathbf{G}(t) \, dt \tag{11}$$

For an isotropic body $\hat{\mathbf{G}}$ is expressed as

$$\hat{G}_{ijkl}(\omega) = \hat{\lambda}(\omega)\delta_{ij}\delta_{kl} + \hat{\mu}(\omega)(\delta_{ik}\delta_{jl} + \delta_{il}\delta_{jk}) \tag{12}$$

where $\hat{\lambda}$ and $\hat{\mu}$ are complex functions of ω. For example, the so-called 'Voigt model with a constant Poisson's ratio', defined by

$$\sigma_{ij} = \lambda \delta_{ij}\varepsilon_{kk} + 2\mu\varepsilon_{ij} + \gamma(\lambda \delta_{ij}\dot{\varepsilon}_{kk} + 2\mu\dot{\varepsilon}_{ij}) \tag{13}$$

gives

$$\hat{\lambda}(\omega) = \lambda(1 - i\omega\gamma) \tag{14a}$$

$$\hat{\mu}(\omega) = \mu(1 - i\omega\gamma) \tag{14b}$$

where γ is a constant. Correspondingly the wave numbers for viscoelastic materials defined by

$$\hat{k}_L = \omega\left(\frac{\rho}{\hat{\lambda} + 2\hat{\mu}}\right)^{1/2} \tag{15a}$$

$$\hat{k}_T = \omega\left(\frac{\rho}{\hat{\mu}}\right)^{1/2} \tag{15b}$$

are complex numbers with positive imaginary parts for $\omega > 0$.

5.3 BIE FORMULATIONS

5.3.1 Conventional BIE Formulations (Kupradze *et al.*, 1979; Kobayashi, 1985)

5.3.1.1 BIE in Fourier Transformed Domain
The reduced problem is easily formulated into boundary integral equations when the material is isotropic and homogeneous.

The integral representation of the solution is expressed by

$$\tilde{\mathbf{u}}(\mathbf{x}; \omega) = \int_{\partial D} \hat{\mathbf{U}}(\mathbf{x}, \mathbf{y}; \omega) \cdot \mathbf{t}(\mathbf{y}; \omega) \, dS_y - \int_{\partial D} \hat{\mathbf{W}}(\mathbf{x}, \mathbf{y}; \omega) \cdot \mathbf{u}(\mathbf{y}; \omega) \, dS_y \tag{16}$$

where

$$\tilde{\mathbf{u}} = \begin{cases} \mathbf{u} & \text{in } D \\ 0 & \text{in } D^c \backslash \partial D \end{cases} \tag{17}$$

$\hat{\mathbf{U}}$ is the fundamental solution of eqn (5) given by

$$\hat{U}_{ij} = \frac{1}{4\pi\mu}\left[\frac{e^{ik_T r}}{r}\delta_{ij} + \frac{1}{k_T^2}\partial_i\partial_j\left(\frac{e^{ik_T r}}{r} - \frac{e^{ik_L r}}{r}\right)\right] \tag{18}$$

where

$$r := |\mathbf{x} - \mathbf{y}| \tag{19a}$$

$$k_{L,T} = \omega/c_{L,T} \tag{19b}$$

$$c_L = \left(\frac{\lambda + 2\mu}{\rho}\right)^{1/2} \tag{19c}$$

$$c_T = \left(\frac{\mu}{\rho}\right)^{1/2} \tag{19d}$$

$\hat{\mathbf{W}} = (\overset{n}{\mathbf{T}}\hat{\mathbf{U}})^T$, and D^c denotes the complement of D, respectively.

Taking the limit by letting the observation point tend to the boundary, we obtain the boundary integral equation given by

$$\mathbf{N} \cdot \mathbf{u}(\mathbf{x}; \omega) = \int_{\partial D} \hat{\mathbf{U}}(\mathbf{x}, \mathbf{y}; \omega) \cdot \mathbf{t}(\mathbf{y}; \omega) \, dS_y - \oint_{\partial D} \hat{\mathbf{W}}(\mathbf{x}, \mathbf{y}; \omega) \cdot \mathbf{u}(\mathbf{y}; \omega) \, dS_y, \quad \mathbf{x} \in \partial D \tag{20}$$

where $\mathbf{N}$ stands for the non-integral term of the double layer potential, and $\oint$ indicates an integral in the sense of Cauchy's principal value. The non-integral term $\mathbf{N} \cdot \mathbf{u}$ is equal to $\mathbf{u}/2$ on the smooth part of the boundary. In exterior problems, one may modify eqn (16) by adding the incident displacement field $\mathbf{u}_I$ to its right-hand side.

5.3.1.2 BIE in Time Domain
A BIE for isotropic homogeneous materials in the time domain is obtained as the Fourier inversion of eqn (20). With the use of the convolution integral defined by

$$\mathbf{a}(\mathbf{x}, t) * \mathbf{b}(\mathbf{x}, t) = \int_0^t \mathbf{a}(\mathbf{x}, \tau) \cdot \mathbf{b}(\mathbf{x}, t - \tau) \, d\tau \tag{21}$$

we can obtain the BIEM in the time domain simply by replacing the dot products on the right-hand side of eqn (20) by convolutions, and kernel functions by their time domain counterparts, respectively.

5.3.2 Domain-typed Integral Equations
It is of practical importance to obtain solutions of inhomogeneous anisotropic elastodynamic problems, because of many earthquake engin-

eering applications of this type such as structure–foundation interactions, etc. For this purpose, we here formulate boundary domain integral equations (Kitahara *et al.*, 1984; Niwa *et al.*, 1986; Hirose and Kitahara, 1989) for inhomogeneous anisotropic elastodynamics. The difficulty for this problem, however, is that no fundamental solutions are available even in homogeneous anisotropic elastodynamics or in inhomogeneous isotropic elastodynamics. To circumvent this difficulty, we use the fundamental solution for homogeneous isotropic elastostatics. In this case, the effects of inhomogeneity, anisotropy and acceleration are all considered with the help of volume potentials. For boundary-typed formulations in anisotropic elasticity, we refer to Rizzo and Shippy (1970), Vogel and Rizzo (1973), Wilson and Cruse (1978), and Nishimura and Kobayashi (1983) for homogeneous anisotropic elastostatic BIEM, and to Nishimura *et al.* (1986) for homogeneous anisotropic elastodynamics.

5.3.2.1 Preliminaries

We consider the reduced elastodynamic problem for non-constant $C_{ijkl}(\mathbf{x})$; the time-domain solution is obtained by using the FFT algorithm.

The basic idea in formulating the integral equation for eqn (8) is to use the fundamental solution for homogeneous elastostatics defined as

$$\nabla \cdot (\mathbf{C^*} : \nabla \mathbf{U^*}(\mathbf{x}, \mathbf{y})) = -\mathbf{1}\delta(\mathbf{x} - \mathbf{y}) \tag{22}$$

where $\mathbf{1}$ is the unit tensor. The constant elasticity tensor C^*_{ijkl} in eqn (22) is arbitrary. However, in an actual calculation, C^*_{ijkl} will be chosen as isotropic (see eqn (4)) with no loss of generality. In this case, the fundamental solution $\mathbf{U^*}$ has the well-known form

$$U^*_{im}(\mathbf{x}, \mathbf{y}) = \frac{1}{8\pi(\lambda^* + 2\mu^*)r}\{(\lambda^* + 3\mu^*)\delta_{im} + (\lambda^* + \mu^*)r_{,i}r_{,m}\} \tag{23}$$

for three-dimensional problems and

$$U^*_{\alpha\sigma}(\mathbf{x}, \mathbf{y}) = \frac{1}{4\pi\mu^*(\lambda^* + 2\mu^*)}\left[(\lambda^* + 3\mu^*)\delta_{\alpha\sigma}\log\frac{1}{r} + (\lambda^* + \mu^*)r_{,\alpha}r_{,\sigma}\right] \tag{24}$$

with

$$U^*_{33}(\mathbf{x}, \mathbf{y}) = \frac{1}{2\pi\mu^*}\log\frac{1}{r} \tag{25}$$

for two-dimensional problems. In these formulae λ^* and μ^* are Lame's constants associated with $\mathbf{C^*}$. One may use arbitrary values for these constants in the ensuing analysis. The mean values of λ and μ in D, for example, may be used in the analysis for non-homogeneous isotropic cases. Hereafter, Roman indices run from 1 to 3 and Greek indices from 1 to 2.

5.3.2.2 Potentials
We here cite from Nishimura and Kobayashi (1987) several formulae for anisotropic elastic potentials, which will play an important role in our formulation.

5.3.2.2 (a) Surface potentials. The simple layer potential

$$V_i^*(\mathbf{x}) = \int_{\partial D} U_{im}^*(\mathbf{x}, \mathbf{y}) \varphi_m(\mathbf{y}) \, dS_y \tag{26}$$

is continuous everywhere in R^N ($N = 2, 3$).

The 'generalised double layer' potential is defined as

$$W_i^*(\mathbf{x}) = \int_{\partial D} U_{im,j}^*(\mathbf{x}, \mathbf{y}) \psi_{mj}(\mathbf{y}) \, dS_y \tag{27}$$

where ψ_{mj} is a sufficiently smooth density. The comma and subscript $,j$ denote $\partial/\partial y_j$. This potential has a finite limit $W_i^{*+}(W_i^{*-})$ given by

$$W_i^{*\pm} = \pm \tfrac{1}{2} n_j \Delta_{im}^{*-1}(\mathbf{n}) \psi_{mj}(\mathbf{x}) + \int_{\partial D} U_{im,j}^*(\mathbf{x}, \mathbf{y}) \psi_{mj}(\mathbf{y}) \, dS_y, \quad \mathbf{x} \in \partial D \tag{28}$$

as the field point $\mathbf{x}$ approaches ∂D from the side into which the positive (negative) normal vector points. In eqn (28) $\Delta_{ij}^{*-1}(\mathbf{n})$ stands for the inverse to $\Delta_{ij}^*(\mathbf{n}) = C_{ikjl}^* n_k n_l$. In the isotropic case we have

$$\Delta_{im}^{*-1}(\mathbf{n}) = \frac{1}{\mu^*} \left(\delta_{im} - \frac{\lambda^* + \mu^*}{\lambda^* + 2\mu^*} n_i n_m \right) \tag{29}$$

5.3.2.2 (b) Volume potentials. The volume potentials

$$D_i^*(\mathbf{x}) = \int_D U_{im}^*(\mathbf{x}, \mathbf{y}) \varphi_m(\mathbf{y}) \, dV_y \tag{30}$$

$$D_i'^*(\mathbf{x}) = \int_D U_{im,j}^*(\mathbf{x}, \mathbf{y}) \psi_{mj}(\mathbf{y}) \, dV_y \tag{31}$$

are continuous everywhere in R^N.

The volume potential defined as

$$D''^*_i(\mathbf{x}) = \int_D U^*_{im,jk}(\mathbf{x}, \mathbf{y}) \Psi_{mjk}(\mathbf{y}) \, dV_y \tag{32}$$

is written explicitly as

$$D''^*_i(\mathbf{x}) = \begin{cases} -v_{kimj}\Psi_{mjk}(\mathbf{x}) + \displaystyle\int_D U^*_{im,jk}(\mathbf{x}, \mathbf{y})\Psi_{mjk}(\mathbf{y}) \, dV_y, & \mathbf{x} \in D \\[3mm] \displaystyle\int_D U^*_{im,jk}(\mathbf{x}, \mathbf{y})\Psi_{mjk}(\mathbf{y}) \, dV_y, & \mathbf{x} \in D^c \backslash \partial D \end{cases} \tag{33}$$

where v_{ijkl} is a tensor defined as

$$v_{ijkl} = \frac{1}{|S^N|} \int_{S^N} \xi_i \Delta_{jk}^{*-1}(\xi) \xi_l \, dS \tag{34}$$

S^N is the unit sphere in R^N, and $|S^N|$ is the area of S^N, respectively. The boundary values of $D''^*_i(\mathbf{x})$ on ∂D have the following expression:

$$D''^{*\pm}_i = -\tfrac{1}{2}v_{kimj}\Psi_{mjk} \pm \tfrac{1}{2}n_k\Delta_{im}^{*-1}(\mathbf{n})n_j\Psi_{mjk}(\mathbf{x})$$

$$+ \int_{\partial D} U^*_{im,jk}(\mathbf{x}, \mathbf{y})\Psi_{mjk}(\mathbf{y}) \, dV_y, \quad \mathbf{x} \in \partial D \tag{35}$$

In the isotropic case, v_{ijkl} in eqn (34) has an explicit form given by

$$v_{ijkl} = \frac{\{N(\lambda^* + 2\mu^*) + (\lambda^* + 3\mu^*)\}\delta_{il}\delta_{jk} - (\lambda^* + \mu^*)(\delta_{ij}\delta_{kl} + \delta_{ik}\delta_{jl})}{\mu^*(\lambda^* + 2\mu^*)N(N+2)} \tag{36}$$

5.3.2.3 Integral Equations
The integral equations for inhomogeneous anisotropic elastodynamics are derived here with the help of the fundamental solution for homogeneous isotropic elastostatics given in eqns (23)–(25). Namely, we use eqn (4) for C^*_{ijkl} with (λ, μ) replaced by (λ^*, μ^*). The elastic moduli $C_{ijkl}(\mathbf{x})$ are assumed to be sufficiently smooth in D.

5.3.2.3 (a) Three-dimensional problem. We shall start with

$$0 = \int_D U^*_{mi}(\mathbf{x}, \mathbf{y})\{(C_{ijkl}(\mathbf{y})u_{k,l}(\mathbf{y}))_{,j} + \rho b_i(\mathbf{y}) + \rho\omega^2 u_i(\mathbf{y})\} \, dV_y \tag{37}$$

which follows from eqn (8), where $\mathbf{U}^*$ is defined in eqn (22). Using the divergence theorem twice, we rewrite eqn (37) into

$$\int_{\partial D} \{U^*_{mi}(\mathbf{x}, \mathbf{y})t_i(\mathbf{y}) - W^*_{mi}(\mathbf{x}, \mathbf{y})u_i(\mathbf{y})\}\, dS_y + \int_D (U^*_{mi,j}(\mathbf{x}, \mathbf{y})C_{ijkl}(\mathbf{y}))_{,l}u_k(\mathbf{y})\, dV_y$$

$$+ \int_D U^*_{mi}(\mathbf{x}, \mathbf{y})(\rho(\mathbf{y})b_i(\mathbf{y}) + \rho(\mathbf{y})\omega^2 u_i(\mathbf{y}))\, dV_y = 0, \quad \mathbf{x} \in R^N \backslash \partial D \tag{38}$$

where $t_i(\mathbf{y})$ is expressed as

$$t_i(\mathbf{y}) = n_j(\mathbf{y})C_{ijkl}(\mathbf{y})u_{k,l}(\mathbf{y}) \tag{39}$$

and $W^*_{mi}(\mathbf{x}, \mathbf{y})$ is defined by

$$W^*_{mi}(\mathbf{x}, \mathbf{y}) = n_l(\mathbf{y})C_{kjil}(\mathbf{y})U^*_{mk,j}(\mathbf{x}, \mathbf{y}) \tag{40}$$

The differentiation in the second integral in eqn (38) is in the sense of distribution. Using the properties of potentials shown in Section 3.2.2 we obtain the following integral equations defined in D and on ∂D:

$$\int_{\partial D} \{U^*_{mi}(\mathbf{x}, \mathbf{y})t_i(\mathbf{y}) - W^*_{mi}(\mathbf{x}, \mathbf{y})u_i(\mathbf{y})\}\, dS_y + \fint_D U^*_{mi,ji}(\mathbf{x}, \mathbf{y})C_{ijkl}(\mathbf{y})u_k(\mathbf{y})\, dV_y$$

$$+ \int_D U^*_{mi,j}(\mathbf{x}, \mathbf{y})C_{ijkl,l}(\mathbf{y})u_k(\mathbf{y})\, dV_y + \int_D U^*_{mi}(\mathbf{x}, \mathbf{y})\rho(\mathbf{y})b_i(\mathbf{y})\, dV_y$$

$$+ \omega^2 \int_D U^*_{mi}(\mathbf{x}, \mathbf{y})\rho(\mathbf{y})u_i(\mathbf{y})\, dV_y = v_{lmij}C_{ijkl}(\mathbf{x})u_k(\mathbf{x}), \quad \mathbf{x} \in D \tag{41}$$

$$\int_{\partial D} U^*_{mi}(\mathbf{x}, \mathbf{y})t_i(\mathbf{y})\, dS_y - \fint_{\partial D} W^*_{mi}(\mathbf{x}, \mathbf{y})u_i(\mathbf{y})\, dS_y$$

$$+ \fint_D U^*_{mi,ji}(\mathbf{x}, \mathbf{y})C_{ijkl}(\mathbf{y})u_k(\mathbf{y})\, dV_y + \int_D U^*_{mi,j}(\mathbf{x}, \mathbf{y})C_{ijkl,l}(\mathbf{y})u_k(\mathbf{y})\, dV_y$$

$$+ \int_D U^*_{mi}(\mathbf{x}, \mathbf{y})\rho(\mathbf{y})b_i(\mathbf{y})\, dV_y + \omega^2 \int_D U^*_{mi}(\mathbf{x}, \mathbf{y})\rho(\mathbf{y})u_i(\mathbf{y})\, dV_y$$

$$= \tfrac{1}{2}v_{lmij}C_{ijkl}(\mathbf{x})u_k(\mathbf{x}), \quad \mathbf{x} \in \partial D \tag{42}$$

Equations (41) and (42) are boundary-domain integral equations which have u_i in D, and u_i and t_i on ∂D as unknowns. It is to be remarked that, even with the isotropic fundamental solution U^*, the effect of anisotropy is included in the C_{ijkl} terms in eqns (41) and (42). The kernel $W^*_{mi}(x, y)$ also includes the anisotropic effect as we can see in eqn (40).

5.3.2.3 (b) Two-dimensional problem. The two-dimensional version of the foregoing analysis is obtained in a similar manner. Actually, we have

$$\int_{\partial D} \{U^*_{\sigma\alpha}(\mathbf{x}, \mathbf{y})t_\alpha(\mathbf{y}) - W^*_{\sigma k}(\mathbf{x}, \mathbf{y})u_k(\mathbf{y})\}\, dS_y + \int_{D} U^*_{\sigma\alpha,\beta\gamma}(\mathbf{x}, \mathbf{y})C_{\alpha\beta k\gamma}(\mathbf{y})u_k(\mathbf{y})\, dV_y$$

$$+ \int_{D} U^*_{\sigma\alpha,\beta}(\mathbf{x}, \mathbf{y})C_{\alpha\beta k\gamma,\gamma}(\mathbf{y})u_k(\mathbf{y})\, dV_y + \int_{D} U^*_{\sigma\alpha}(\mathbf{x}, \mathbf{y})\rho(\mathbf{y})b_\alpha(\mathbf{y})\, dV_y$$

$$+ \omega^2 \int_{D} U^*_{\sigma\alpha}(\mathbf{x}, \mathbf{y})\rho(\mathbf{y})u_\alpha(\mathbf{y})\, dV_y$$

$$= \begin{bmatrix} v_{\gamma\sigma\alpha\beta}C_{\alpha\beta k\gamma}(\mathbf{x})u_k(\mathbf{x}), & \mathbf{x} \in D \\ \tfrac{1}{2}v_{\gamma\sigma\alpha\beta}C_{\alpha\beta k\gamma}(\mathbf{x})u_k(\mathbf{x}), & \mathbf{x} \in \partial D \\ 0, & \mathbf{x} \in D^c \backslash \partial D \end{bmatrix} \qquad (43)$$

$$\int_{\partial D} \{U^*_{33}(\mathbf{x}, \mathbf{y})t_3(\mathbf{y}) - W^*_{3k}(\mathbf{x}, \mathbf{y})u_k(\mathbf{y})\}\, dS_y + \int_{D} U^*_{33,\beta\gamma}(\mathbf{x}, \mathbf{y})C_{3\beta k\gamma}(\mathbf{y})u_k(\mathbf{y})\, dV_y$$

$$+ \int_{D} U^*_{33,\beta}(\mathbf{x}, \mathbf{y})C_{3\beta k\gamma,\gamma}(\mathbf{y})u_k(\mathbf{y})\, dV_y + \int_{D} U^*_{33}(\mathbf{x}, \mathbf{y})\rho(\mathbf{y})b_3(\mathbf{y})\, dV_y$$

$$+ \omega^2 \int_{D} U^*_{33}(\mathbf{x}, \mathbf{y})\rho(\mathbf{y})u_3(\mathbf{y})\, dV_y$$

$$= \begin{bmatrix} \dfrac{1}{2\mu^*} C_{3\beta k\beta}(\mathbf{x})u_k(\mathbf{x}), & \mathbf{x} \in D \\ \dfrac{1}{4\mu^*} C_{3\beta k\beta}(\mathbf{x})u_k(\mathbf{x}), & \mathbf{x} \in \partial D \\ 0, & \mathbf{x} \in D^c \backslash \partial D \end{bmatrix} \qquad (44)$$

where $W^*_{\alpha k}$ and W^*_{3k} are defined as

$$W^*_{\alpha k}(\mathbf{x}, \mathbf{y}) = n_\gamma(\mathbf{y})C_{\alpha\beta k\gamma}(\mathbf{y})U^*_{\sigma\alpha,\beta}(\mathbf{x}, \mathbf{y}) \tag{45a}$$

$$W^*_{3k}(\mathbf{x}, \mathbf{y}) = n_\gamma(\mathbf{y})C_{3\beta k\gamma}(\mathbf{y})U^*_{33,\beta}(\mathbf{x}, \mathbf{y}) \tag{45b}$$

The v_{ijkl} in eqn (43) are defined in eqn (34) and have the form given in eqn (36) with $N = 2$. In eqn (44) we have used

$$v_{\gamma 33\beta} = \frac{1}{2\mu^*}\delta_{\gamma\beta} \tag{46}$$

which follows from eqn (34) and

$$\Delta^{*-1}_{33} = \frac{1}{\mu^*(\xi_1^2 + \xi_2^2)} \tag{47}$$

Note that the double layer potentials in eqns (43) and (44) are evaluated in the sense of Cauchy's principal value when $\mathbf{x} \in \partial D$. From eqns (43) and (44), we see that the in-plane and anti-plane motions are not independent for the general anisotropic body. For the special case of orthotropy, however, eqns (43) and (44) decouple if the principal axes of anisotropy coincide with the coordinate axes.

5.3.3 Regularisation of Hypersingular Kernels in BIE

Another effective application of BIEM is found in problems including singularities. This is because the BIEM constructs approximate solutions by superposing singular solutions. In particular, BIEM is suitable for crack problems because the double layer potential determines an elastodynamic displacement field with a given displacement discontinuity in an automatic manner. In the double layer BIE formulations, however, one encounters integral equations with hypersingular kernels, which are difficult to handle numerically. Hence in this section we shall discuss a method for reducing these singular integrals to ordinary integrals. For related topics the reader is referred to Sládek and Sládek (1984), Nedelec (1986) and Polch et al. (1987).

To fix the idea, we consider 2D time-domain BIEMs for crack problems assuming that the material is isotropic and homogeneous. Let D be $R^2 \backslash S$ where S is a smooth curve of finite length. Also, let $\mathbf{n}$ be a unit normal vector to S, which points into one and the same side of S called the positive side. Our problem is to find a solution $\mathbf{u}(\mathbf{x}, t)$ of eqn (5) subject to initial conditions

$$\mathbf{u}|_{t=0} = \mathbf{u}_1|_{t=0} \tag{48a}$$

$$\dot{\mathbf{u}}|_{t=0} = \dot{\mathbf{u}}_1|_{t=0} \tag{48b}$$

boundary condition

$$\overset{n}{T}u^{\pm} = 0 \text{ on } S \times T \qquad (49)$$

radiation condition

$$u = u_I \quad \text{for } |x| > ct + d \quad (c, d = \text{const.}) \qquad (50)$$

and regularity condition

$$\lim_{x \to x_0} [u(x)] := \lim_{x \to x_0} (u^+(x) - u^-(x)) = 0, \quad x \in S, \quad x_0 \in \partial S \qquad (51)$$

where u_I stands for an incident field which satisfies eqn (5) in $R^2 \times T$ identically.

As is already known, the solution to this problem is written as

$$u(x, t) = u_I(x, t) + \int_0^t \int_S W(x, y, t - s) \cdot \varphi(y, s) \, dS \, ds \qquad (52)$$

where $\varphi := u^+ - u^-$ and W stands for the double layer kernel defined in terms of the fundamental solution U as

$$W_{il}(x, y, t) := \frac{\partial}{\partial y_k} U_{ij}(x - y, t) C_{jklm} n_m(y) \qquad (53)$$

The time domain fundamental solution U has a well-known form

$$U_{ij}(x, t) = \delta_{ij}(F + G) + \frac{x_i x_j}{|x|} \frac{\partial}{\partial |x|} G \qquad (54)$$

where

$$F(|x|, t) = \frac{1}{2\pi \rho c_T \{(c_T t)^2 - |x|^2\}_+^{1/2}} \qquad (55)$$

$$G(|x|, t) = \frac{1}{2\pi \rho |x|^2} \left(\frac{\{(c_T t)^2 - |x|^2\}_+^{1/2}}{c_T} - \frac{\{(c_L t)^2 - |x|^2\}_+^{1/2}}{c_L} \right) \qquad (56)$$

and $\{x\}_+^{1/2}$ is defined by

$$\{x\}_+^{1/2} := \begin{cases} x^{1/2} & \text{if } x > 0 \\ 0 & \text{otherwise} \end{cases} \qquad (57)$$

The integral in eqn (52) is taken in the sense of the finite part on the past cones where either $|x - y| = c_T(t - s)$ or $|x - y| = c_L(t - s)$ holds.

From eqns (49) and (52) follows a boundary-time 'integral' equation given by

$$0 = \overset{n}{T}\mathbf{u}_I + pf \int_0^t \int_S \overset{n}{T}\mathbf{W} \cdot \boldsymbol{\varphi} \, dS \, ds \tag{58}$$

The solution to this equation determines $\mathbf{u}$ in $D \times T$ completely through eqn (52).

It is seen that the kernel function $\overset{n}{T}\mathbf{W}$ has a singularity at the observation point $(\mathbf{x} = \mathbf{y})$. The order of singularity is $1/|\mathbf{x} - \mathbf{y}|^2$ as one can show after performing the time integration. Also, on a past cone defined by $|\mathbf{x} - \mathbf{y}| = c(t - s)$ for fixed $\mathbf{x}$ and t, eqn (58), as an integral with respect to $(\mathbf{y}, s)$, has a singularity of order $(c|t - s| - |\mathbf{x} - \mathbf{y})^{-5/2}$, where c is either c_L or c_T. All the integrals in eqn (58) including these singularities are to be understood in the sense of the finite part.

From a numerical analytical point of view, eqn (58) is not desirable because of these hypersingular integrals. As shown in Guo et al. (1987), however, manipulations similar to those used in Nishimura et al. (1987) transform eqn (58) into

$$-(\overset{n}{T}u_I(\mathbf{x}, t))_s = C_{sqip} n_q(\mathbf{x}) \left[-\rho \int_S \int_0^t U_{im} n_p \frac{\partial^2 \varphi_m}{\partial s^2} \, ds \, dS_y \right.$$

$$- pf \int_S \int_0^t \varepsilon_{pl} [U_{ij} + (s_i n_j + s_j n_i) G] C_{jklm} \left(\kappa n_k \frac{\partial \varphi_m}{\partial \xi} + s_k \frac{\partial^2 \varphi_m}{\partial \xi^2} \right) ds \, dS_y$$

$$- \int_S \int_0^t \varepsilon_{pl} n_k C_{jklm} \left[\left\{ 2G\kappa(n_i n_j - s_i s_j) - \left(\delta_{ij} + 2n_i n_j - \frac{4(x_i - y_i)(x_j - y_j)}{|\mathbf{x} - \mathbf{y}|^2} \right) \frac{\partial G}{\partial n_y} \right\} \frac{\partial \varphi_m}{\partial \xi} \right.$$

$$+ \left. \left(\frac{\partial |\mathbf{x} - \mathbf{y}|}{\partial n_y} \delta_{ij} F + \frac{(x_i - y_i)(x_j - y_j)}{|\mathbf{x} - \mathbf{y}|^3} \frac{\partial}{\partial n_y} (|\mathbf{x} - \mathbf{y}|^2 G) \right) \frac{t - s}{|\mathbf{x} - \mathbf{y}|} \frac{\partial^2 \varphi_m}{\partial \xi \partial s} \right] ds \, dS_y \right] \tag{59}$$

where $\partial/\partial\xi$, ε_{ij}, κ, pf and s stand for the tangential differentiation along S, permutation symbol, the curvature of S, finite part, and the unit tangent vector to S at $\mathbf{y}$, respectively. The pf symbol is due to the well-known square root singularity of φ at the tip. Note that all the integrals in eqn (59) are convergent in the ordinary manner except for the pf integrals of the following form:

$$I = \mathrm{pf} \int_0^t \int_S K(\mathbf{x}, \mathbf{y}, t-s) \frac{\partial^2 \varphi}{\partial \xi^2}(\mathbf{y}, s)\, \mathrm{d}S_y\, \mathrm{d}s \tag{60}$$

where K is a certain kernel function whose singularity is integrable.

In numerical calculation one usually discretises BIEs by introducing some shape functions. A natural set of shape functions in the present application is obtained as a linear combination of functions of the following form:

$$\psi(\mathbf{y}) f(s) \tag{61}$$

where ψ is a function of $\mathbf{y}$ and f is a function of s, respectively. With this choice of the shape functions we can write I in terms of ordinary integrals. Indeed, after carrying out the time integration in eqn (60) analytically with eqn (61), one obtains an expression for I in the following form:

$$I = \mathrm{pf} \int_S K'(\mathbf{x}, \mathbf{y}, t-s) \frac{\partial^2 \psi}{\partial \xi^2}(\mathbf{y})\, \mathrm{d}S_y \tag{62}$$

where K' is a certain kernel function having an integrable singularity at $\mathbf{x} = \mathbf{y}$. Hence we have

$$I = K'(\mathbf{x}, \mathbf{y}_0)\mathrm{pf} \int_S \frac{\partial^2 \psi}{\partial \xi^2}\, \mathrm{d}S_y + \int_S (K'(\mathbf{x}, \mathbf{y}) - K'(\mathbf{x}, \mathbf{y}_0)) \frac{\partial^2 \psi}{\partial \xi^2}\, \mathrm{d}S_y \tag{63}$$

where $\mathbf{y}_0$ is the location of a tip. Note that the first integral in eqn (62) admits an analytical integration for a particular choice of ψ because this term is an integral of a tangential derivative. In addition the last integral in eqn (63) is convergent at $\mathbf{y}_0$. Therefore eqn (63) gives an expression for eqn (62) in terms of integrable integrals. Now that all the integrals in eqn (59) are convergent in the ordinary sense, we can apply the conventional BIEM approach to the integral equation in eqn (58) with the help of eqn (59).

5.4 NUMERICAL TECHNIQUES

5.4.1 Combination of BIEM and FEM (Kobayashi and Mori, 1986; Kobayashi et al., 1986)

It is well recognised that the finite element method (FEM) is effective for interior problems which include inhomogeneity and/or non-linearity, and that the boundary integral equation method (BIEM) is most advantageously applied to linear problems for unbounded domains. Therefore, the combination of them (BIE–FE hybrid method) is expected to be ideal for those wave problems in which the inhomogeneity and non-linearity are important only in a bounded region.

In practical applications, we often have to deal with various types of structures constructed on or in non-homogeneous and non-linear soils. In analysing these kinds of structure–soil system, the combination of BIEM and FEM is considered to be effective. Namely, we use FEM for the interior domain which has inhomogeneous and/or non-linear properties, and the BIEM for the surrounding infinite or semi-infinite domain which is well regarded as homogeneous and isotropic.

In order to fix the idea, we consider a model whose interior domain is embedded in an exterior half-space. We denote quantities relevant to the interior domain, the exterior domain, the exposed boundary of the interior domain, the exposed boundary of the exterior domain, and the common boundary of the interior and exterior domains (see Fig. 1) by subscripts f, b, s_i, s_e and s, respectively. We also assume that traction is given on the exposed boundaries.

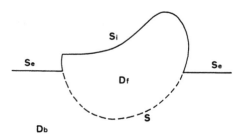

FIG. 1   Domains and boundaries.

The FE equation for the reduced elastodynamics is expressed in terms of nodal displacement **d** by

$$\mathbf{K} \cdot \mathbf{d} = \mathbf{f} \tag{64}$$

where **K** and **f** stand for the stiffness matrix and the nodal force vector defined by

$$\mathbf{K} = \int_{D} \mathbf{B}^{\mathrm{T}} \mathbf{D} \mathbf{B} \, \mathrm{d}V - \rho \omega^2 \int_{D} \mathbf{N}^{\mathrm{T}} \mathbf{N} \, \mathrm{d}V \tag{65a}$$

$$\mathbf{f} = \int_{\partial D} \mathbf{N}^T \mathbf{N} \cdot \mathbf{t} \, dS \tag{65b}$$

respectively. In eqns (65), $\mathbf{N}$ and $\mathbf{N}^T$ stand for the matrix of shape functions and its transpose, and $\mathbf{B}$ and $\mathbf{D}$ are the strain matrix and elasticity matrix, respectively.

The BIE for the reduced problem is discretised as usual by using, say, N M-node isoparametric elements. Indeed, the use of the local coordinate $(0, \xi_1, \xi_2)$ gives

$$N_{ki}(\mathbf{x}^\beta) u_i(\mathbf{x}^\beta; \omega) + \sum_{n=1}^{N} \sum_{\alpha=1}^{M} W_{kin}^{\beta\alpha} u_i(\mathbf{x}^{(n,\alpha)}; \omega)$$

$$= u_{1k}(\mathbf{x}^\beta; \omega) + \sum_{n=1}^{N} \sum_{\alpha=1}^{M} U_{kin}^{\beta\alpha} t_i(\mathbf{x}^{(n,\alpha)}; \omega) \tag{66}$$

where $\mathbf{x}^\beta$ is the βth nodal point,

$$W_{kin}^{\beta\alpha} := \int_{-1}^{1} \int_{-1}^{1} \hat{W}_{ki}(\mathbf{x}^\beta, \mathbf{y}(\xi); \omega) N^\alpha(\xi) J_n(\xi) \, d\xi_1 \, d\xi_2 \tag{67a}$$

$$U_{kin}^{\beta\alpha} := \int_{-1}^{1} \int_{-1}^{1} \hat{U}_{ki}(\mathbf{x}^\beta, \mathbf{y}(\xi); \omega) N^\alpha(\xi) J_n(\xi) \, d\xi_1 \, d\xi_2 \tag{67b}$$

$J_n(\xi)$ is the Jacobian of the isoparametric mapping from $(-1, 1) \times (-1, 1)$ to the nth element, and (n, α) stands for the global node number of the αth node of the nth element, respectively. Also, the summation rule is used for the index i. We then rearrange eqn (64) by using the subscripts b and f mentioned above to obtain

$$[\mathbf{W}_{s_e}, \mathbf{W}_s] \begin{bmatrix} \mathbf{u}_{s_e} \\ \mathbf{u}_{s_b} \end{bmatrix} = [\mathbf{u}_I] + [\mathbf{U}_{s_e}, \mathbf{U}_s] \begin{bmatrix} \mathbf{t}_{s_e} \\ \mathbf{t}_{s_b} \end{bmatrix} \tag{68}$$

where we have included the contribution from $N_{ki}(\mathbf{x}^\beta)$ in eqn (66) into $[\mathbf{W}_{s_e}, \mathbf{W}_s]$.

We now combine the FE and the BI equations. Using the boundary conditions and the connectivity conditions given by

$$\mathbf{t}_{s_e} = \bar{\mathbf{t}}_{s_e}, \quad \mathbf{t}_{s_i} = \bar{\mathbf{t}}_{s_i}, \quad \mathbf{u}_{s_f} = \mathbf{u}_{s_b}, \quad \mathbf{t}_{s_f} + \mathbf{t}_{s_b} = 0 \tag{69a–d}$$

we have the following system of equations:

$$
\begin{bmatrix}
\mathbf{K}_{ff}, & \mathbf{K}_{fs_i}, & \mathbf{K}_{fs}, & 0, & 0 \\
\mathbf{K}_{s_if}, & \mathbf{K}_{s_is_i}, & \mathbf{K}_{s_is}, & 0, & 0 \\
\mathbf{K}_{sf}, & \mathbf{K}_{ss_i}, & \mathbf{K}_{ss}, & \mathbf{M}, & 0 \\
0, & 0, & \mathbf{W}_s, & -\mathbf{U}_s, & \mathbf{W}_{s_e}
\end{bmatrix}
\begin{bmatrix}
\mathbf{u}_f \\
\mathbf{u}_{s_i} \\
\mathbf{u}_s \\
\mathbf{t}_{s_b} \\
\mathbf{u}_{s_b}
\end{bmatrix}
\begin{bmatrix}
0 \\
\mathbf{f}_{s_i} \\
0 \\
\mathbf{u}_I + \mathbf{U}_{s_e}\bar{\mathbf{t}}_{s_e}
\end{bmatrix}
\tag{70}
$$

where

$$
\mathbf{M} = \int_S \mathbf{N}^T \mathbf{N}\, dS
\tag{71a}
$$

$$
\mathbf{f}_{s_i} = \int_{S_i} \mathbf{N}^T \bar{\mathbf{t}}_{s_i}\, dS
\tag{71b}
$$

respectively.

The right-hand side of eqn (70) is known. Therefore we can solve eqn (70) for the displacement in the interior domain and on the whole boundaries, and for the tractions on the common boundary.

5.4.2 Comments on Discretisation

Discretisation of BIE in the frequency domain has been described in the previous section. The time-domain BIE also admits similar discretisation in space–time (e.g. Kobayashi, 1985).

In crack problems it is preferable to use boundary elements which are square-root singular at tips. In addition, use of C^1 elements is expected to improve the accuracy of the solution by eliminating the undesirable singularities of tractions on the crack face (Nishimura and Kobayashi, 1989). Experience tells, however, that the weighted piecewise constant elements, to be discussed in Section 5.5.3, are not so bad from a practical point of view.

5.4.3 Numerical Fourier Inversion

The FFT algorithm is recommended for the Fourier inversion of the solutions obtained in the frequency domain. Remarks on the use of this method have been made in Kobayashi and Nishimura (1982a).

5.4.4 Boundary Stress
Boundary stresses are evaluated accurately by differentiating boundary displacements, as discussed in Kobayashi and Nishimura (1982a). We note that this method is applicable also with lower-order boundary elements, e.g. piecewise constant elements, etc. Indeed, we interpolate the nodal values of the boundary displacements obtained by BIEM with the help of, say, spline functions. We then take the covariant derivatives of the interpolated boundary displacement, and then follow the method of Kobayashi and Nishimura (1982a) to obtain boundary stresses.

5.4.5 Uniqueness of the Solution
Uniqueness of solutions to the integral equations for reduced elastodynamics does not hold at certain frequencies even in exterior problems. Such frequencies are called 'fictitious eigenfrequencies'. The fictitious eigenfrequency problem has been discussed in Kupradze et al. (1979), Kobayashi and Nishimura (1982a, b) and Kitahara (1985). Remedies for this problem can be found in Kobayashi and Nishimura (1982a, b), Jones (1984), Kitahara (1985) and Bencheikh (1986).

It is noteworthy that the solution to the time-harmonic (reduced elastodynamics) version of the integral equation for crack problems (e.g. Nishimura and Kobayashi, 1989) is unique. Also, the time-domain BIEs for elastodynamics possess unique solutions.

5.5 EXAMPLES

5.5.1 BIE–FE Hybrid Method

5.5.1.1 Displacements of Surface Foundation Subject to SV Waves (Kobayashi and Mori, 1986)
Figure 2 shows the horizontal displacements of a rigid massless rectangular foundation (surface foundation) subject to incident time-harmonic SV waves of oblique incidence. The incident wave is polarised in the x_1–x_3 plane, and the incident angle is $14.5°$ as shown in the inset of Fig. 2. Both the foundation and the ground are modelled elastic. The material constants for the foundation, denoted by subscripts f, and the ground, denoted by subscripts b, are chosen as $\mu_f/\mu_b = 500$, $\rho_f/\rho_b = \frac{1}{10}$ and ν_f (Poisson's ratio) = $\nu_b = \frac{1}{3}$. Also the thickness h of the foundation is chosen as $h = b/10$. With these constants, one may view the foundation to be essentially rigid and massless. We have divided the foundation into four 20-node isoparametric

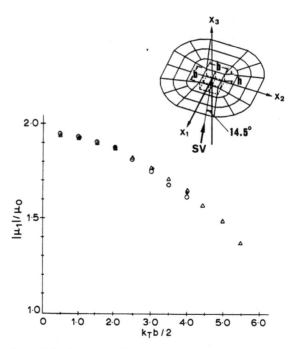

FIG. 2   Horizontal displacement of a rigid massless surface foundation subject to an SV wave of oblique incidence; u_0 = magnitude of the incident wave. ○ Wong and Lucio (1978); △ BIEM (time-domain) (Karabalis *et al.*, 1984); ● BIE–FE.

finite elements, and modelled the ground surface of approximately twice the incident wavelength as shown in the inset of Fig. 2 by using 8-node quadratic boundary elements. Gaussian quadrature formulae are used for the surface and the volume integrals respectively. In spite of the coarse BIEM mesh used, the results obtained by the BIE–FE hybrid method agree very well with those by Wong and Luco (1978) and with the time-domain BIEM results by Karabalis *et al.* (1984).

5.5.1.2 *Response of a T-shaped Bridge Pier Subject to SV Waves (Kobayashi* et al., *1986)*

Figure 3 shows the deformation of an elastic T-shaped bridge pier embedded in a viscoelastic half-space subject to a time-harmonic SV wave of oblique incidence. The model used for the analysis is depicted in Fig. 3(a). The incident wave has a wavelength of L, and is polarised in the vertical plane obtained by rotating the x_1–x_3 plane by 30° around the x_3-axis. Also

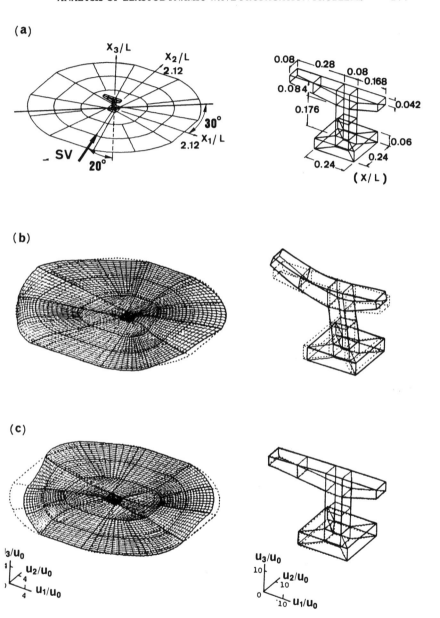

FIG. 3   Deformations of an elastic T-shaped bridge pier–viscoelastic ground system subject to a time-harmonic SV wave of oblique incidence: (a) model used; (b) real part of solution; (c) imaginary part of solution.

the incident angle is $20°$. The material constants for the bridge pier (denoted by subscripts f) and the ground (indicated by subscripts b) are as follows:

$$\hat{\lambda}_b(\omega) = \frac{2\nu_b}{1-2\nu_b}\hat{\mu}_b(\omega) \tag{72a}$$

$$\lambda_f = \frac{2\nu_f}{1-2\nu_f}\mu_f \tag{72b}$$

$$\frac{(1+\nu_b)\hat{\mu}_b(\omega)}{(1+\nu_f)\mu_f} = \frac{1}{100}(1-0.2i) \tag{72c}$$

$$\frac{\rho_b}{\rho_f} = \frac{5}{6}, \quad \nu_b = \frac{1}{3}, \quad \nu_f = \frac{1}{6} \tag{72d–f}$$

See, e.g., eqns (13) and (14) for a physical interpretation of these constants. We have used FEM for the structure and BIEM for the ground. A part of the free surface having a radius of approximately twice the incident wavelength is taken into account in the numerical analysis. The numerical integration is carried out with the help of the same Gaussian quadrature formula as in section 5.5.1.1. The real (imaginary) part of the solution shown in Fig. 3(b) (Fig. 3(c)) indicates the deformation at the moment when the peak (node) of the incident displacement wave reaches the origin of the co-ordinates. The sign of the incident field is chosen such that the real part of the displacement of the incident wave points into the ground at the origin in Fig. 3(b).

The reasonably accurate numerical results in the preceding example motivated the use of essentially the same coarse mesh as in Fig. 2 in the present analysis. For more accurate results, however, it would be safer to refine the mesh.

5.5.2 Domain-typed Integral Equations (Kitahara et al., 1984; Niwa et al., 1986)

All the examples shown here are two-dimensional. Also, the examples of the domain-typed BIEM to follow use $(\lambda^*, \mu^*) = (\lambda^e, \mu^e)$, where λ^e and μ^e are Lame's constants for a certain domain to be specified in each example.

In order to check the accuracy of our present formulation, we first consider a homogeneous isotropic inclusion which has the material properties given by

$$\frac{\lambda}{\lambda^e} = \frac{\mu}{\mu^e} = 0.5, \quad \nu = \nu^e = 0.25, \quad \frac{\rho}{\rho^e} = 1 \tag{73a–c}$$

where (λ, μ, ρ) are the material properties of the inclusion and $(\lambda^e, \mu^e, \rho^e)$ are those of the matrix. In this case, we can also use the conventional BIEM for both inclusion and matrix. Figure 4 compares our present formulation with the conventional BIE formulation. The inclusion is circular with a radius of a. In our present formulation the domain-typed integral equation is used for the inclusion and the conventional BIEM is utilised for the matrix. Two types of incident waves, P and SV, and two wave numbers, $ak_T^e = \pi$ and 2π, were considered in this calculation. k_T^e in Fig. 4 indicates the transverse wave number of the surrounding matrix.

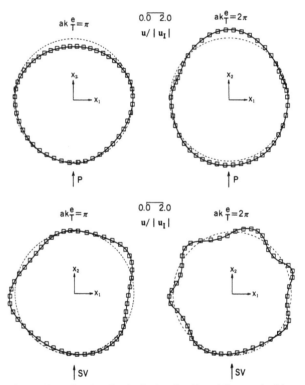

FIG. 4 Deformation of a circular inclusion for P and S wave incidences. ——————: conventional BIEM; □: domain-typed BIEM.

Figure 5 shows the frequency response at point A of an inhomogeneous rectangular structure on a half-space. Poisson's ratio is $\frac{1}{4}$ for the whole system. The ratio of shear modulus $\kappa' = \mu(\mathbf{x})/\mu^e$ varies in the structure as shown in Fig. 5, where μ^e stands for the shear modulus of the homogeneous

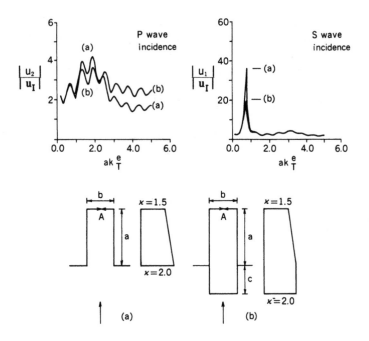

FIG. 5 Frequency response of an inhomogeneous rectangular structure on a half-plane; $\kappa = \mu(x)/\mu^e$, $\nu = \nu^e = 0.25$.

half-space. Two types of structures were considered here. Structure (a) has no embedment, whereas structure (b) has an embedment of length c, where $b/a = c/a = \frac{1}{2}$. The left-hand figure in Fig. 5 gives the response curves for the vertical P wave incidence for two type of structures (a) and (b). u_2 stands for the vertical displacement at points A of the structure. The right-hand figure shows the response curves for the SV wave of vertical incidence. u_1 indicates the horizontal displacement at point A.

Finally we discuss the response of a dam-typed structure to an incident Ricker's SV wavelet. Figure 6 shows an inhomogeneous dam-typed structure on a half-space. The wave form of Ricker's wavelet (Ricker, 1945) with maximum amplitude of A_1 is shown in Fig. 6(a). The angle of incidence of this wavelet, measured from the vertical axis, is α. Material properties are indicated in Fig. 6(b). The shear modulus $\mu(x)$ varies linearly from the dam base to the top surfaces of the core and fillers. Figure 7 shows the transient response of the dam to the Ricker's SV wavelet having an incident angle of 30°.

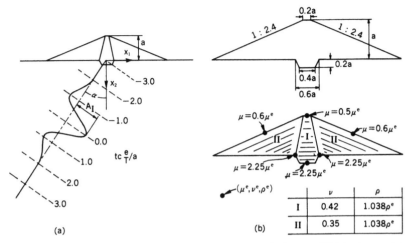

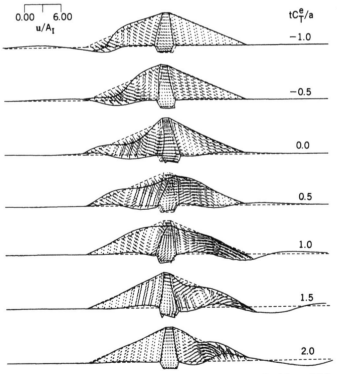

FIG. 6 Inhomogeneous dam-typed structure on a half-plane and Ricker's wavelet; $v^e = 0.28$. (a) Wave form; (b) material properties.

FIG. 7 Transient response of a dam-typed structure to incident Ricker's SV wavelet; α (incident angle) = 30°.

5.5.3 Crack Problems (Guo et al., 1987)

Finally we shall show some numerical examples of the BIE crack analysis. The crack is assumed to be straight, and its length is $2a$. In numerical analysis we introduce a new unknown function φ', in terms of which the crack opening displacement φ is expressed as

$$\varphi(x_1, t) = \sqrt{a^2 - x_1^2}\, \varphi'(x_1, t) \tag{74}$$

We then discretise φ' by using shape functions which are piecewise constant spatially, and piecewise linear in time. Obviously the stress intensity factor is obtained by using the computed values of φ' at the tip.

Figure 8 shows the mode I and II stress intensity factors vs time for the

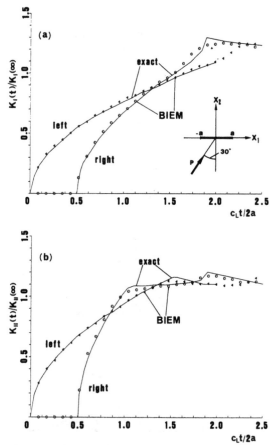

FIG. 8   Stress intensity factors of a straight crack subject to a plane P wave of oblique incidence: (a) K_I; (b) K_{II}.

case of oblique P wave incidence. The incident angle is 30°, the stress magnitude of the incident wave is p_0 (constant), and Poisson's ratio is $\frac{1}{4}$. The BIEM results shown by symbols agree well with the analytical solutions (Thau and Lu, 1971) shown by lines. The corresponding crack opening displacement is plotted in Fig. 9. Again, the solutions obtained by the present method (lines) are in agreement with the frequency-domain BIEM solutions (symbols).

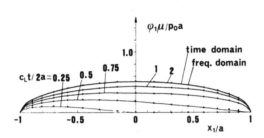

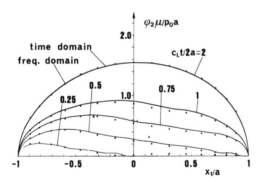

FIG. 9   Crack opening displacement of a straight crack subject to a plane P wave of oblique incidence.

5.6 CONCLUSION

This chapter has presented several possibilities of substructured analysis using BIEM. The combinations of BIEM with FEM or with other BIEM in two- or three-dimensional elastodynamics and viscoelastodynamics have turned out to be successful. Although we did not show examples of the BIE crack analysis combined with other methods, such combination is straightforward with the combination techniques developed for other applications shown in this chapter. Another important application we did not discuss here is the substructured analysis in non-linear dynamic problems. It seems to the present authors, however, that the improvement of stability of existing time-domain BIEMs for elastodynamics is essential to this end.

ACKNOWLEDGEMENTS

The present authors wish to express their gratitude to Mr K. Mori of the Ministry of Construction, to Mr Q.C. Guo of Kyoto University (presently at Northwestern University), and to Dr S. Hirose of Okayama University for their contributions in both numerical and theoretical work.

REFERENCES

BENCHEIKH, L. (1986). Scattering of elastic waves by cylindrical cavities: integral equation methods and low-frequency matched asymptotic expansions, Ph.D. Thesis, University of Manchester, UK.

CHRISTENSEN, R.M. (1971). *Theory of Viscoelasticity: An Introduction*, Academic Press, New York.

ERINGEN, A.C. & SUHUBI, E.S. (1975). *Elastodynamics*, Vol. 2, *Linear theory*, Academic Press, New York.

GUO, Q.C., NISHIMURA, N. & KOBAYASHI, S. (1987). Elastodynamic analysis of a crack by BIEM, *Proc. 4th Japan Nat. Symp. BEM*, Eds S. Kobayashi *et al.*, Japan Assoc. Comp. Mech., Tokyo, 197–202 (in Japanese).

HIROSE, S. & KITAHARA, M. (1989). Elastic wave diffraction of inhomogeneous and anisotropic bodies in a half space, *Earthquake Eng. Struct. Dynamics*, **18**, 285–97.

JONES, D.S. (1984). An exterior problem in elastodynamics, *Math. Proc. Camb. Phil. Soc.*, **96**, 173–82.

KARABALIS, D.L., SPYRAKOS, C.C. & BESKOS, D.E. (1984). Dynamic responses of surface foundations by time domain boundary element method, *Proc. Int. Symp. Dynamic Soil–Structure Interaction*, Eds D.E. Beskos *et al.*, A.A. Balkema, Rotterdam, 19–24.

KITAHARA, M. (1985). *Boundary Integral Equation Methods in Eigenvalue Problems of Elastodynamics and Thin Plate*, Elsevier, Amsterdam.

KITAHARA, M., NIWA, Y., HIROSE, S. & YAMAZAKI, M. (1984). Coupling of numerical Green's matrix and boundary integral equations for the elasto-dynamic analysis of inhomogeneous bodies on an elastic half-space, *Appl. Math. Model.*, **8**, 397–407.

KOBAYASHI, S. (1985). Fundamentals of boundary integral equation methods in elastodynamics, in: *Topics in Boundary Element Research–2*, Ed. C.A. Brebbia, Springer–Verlag, Berlin,1–54.

KOBAYASHI, S. & MORI, K. (1986). Three dimensional dynamic analysis of soil–structure interactions by boundary integral equation–finite element combined method, in: *Innovative Numerical Methods in Engineering*, Eds R.P. Shaw *et al.*, 613–18.

KOBAYASHI, S. & NISHIMURA N. (1982*a*). Transient stress analysis of tunnels and caverns of arbitrary shape due to travelling waves, Chapter 7 in: *Developments in Boundary Element Methods–2*, Eds P.K. Banerjee and R.P. Shaw, Applied Science Publishers, London, 177–210.

KOBAYASHI, S. & NISHIMURA, N. (1982*b*). On the indeterminacy of BIE solutions for the exterior problems of time-harmonic elastodynamics and incompressible elastostatics, *Proc. 4th Int. Conf. BEM*, Ed. C.A. Brebbia, Springer–Verlag, Berlin, 282–96.

KOBAYASHI, S., NISHIMURA, N. & MORI, K. (1986). Applications of boundary element–finite element combined method to three-dimensional viscoelasto-dynamic problems, in: *Boundary Elements*, Ed. Du Quinghua, Pergamon, Oxford, 67–74.

KUPRADZE, V.D., GEGELIA, T.G., BASHELEISHVILI, M.O. & BURCHULADZE, T.V. (1979). *Three-Dimensional Problems of the Mathematical Theory of Elasticity and Thermoelasticity*, North-Holland, Amsterdam.

LACHAT, J.C. & WATSON J.O. (1976). Effective numerical treatment of boundary integral equations: a formulation for three-dimensional elastostatics, *Int. J. Numer. Methods Eng.*, **10**, 991–1005.

NEDELEC, J.C. (1986). The double layer potential for periodic elastic waves in R^3, in: *Boundary Elements*, Ed. Du Quinghua, Pergamon, Oxford, 439–48.

NISHIMURA, N. & KOBAYASHI, S. (1983). A boundary integral equation formulation for three dimensional anisotropic elastostatics, in: *Boundary Elements*, eds C.A. Brebbia *et al.*, Springer–Verlag, Berlin, 345–54.

NISHIMURA, N. & KOBAYASHI, S. (1987). On the behaviour of elastic potentials, *Mem. Fac. Eng., Kyoto Univ.*, **49**, 294–307.

NISHIMURA, N. & KOBAYASHI, S. (1989). A regularized boundary integral equation method for elastodynamic crack problems, *Comput. Mech.*, **4**, 319–28.

NISHIMURA, N. KISHIMA, T. & KOBAYASHI, S. (1986). A BIE analysis of wave propagation in anisotropic media, in: *Boundary Elements VIII*, eds M. Tanaka and C.A. Brebbia, Springer–Verlag, Berlin, 425–34.

NISHIMURA, N. GUO, Q.C. & KOBAYASHI, S. (1987). Boundary integral equation methods in elastodynamic crack problems, in: *Boundary Elements IX*, eds. C.A. Brebbia, W.L. Wendland and G. Kuhn, Springer–Verlag, Berlin, vol. 2, 279–91.

NIWA, Y., HIROSE, S. & KITAHARA, M. (1986). Elastodynamic analysis of inhomo-geneous anisotropic bodies, *Int. J. Solids Struct.*, **22**, 1541–55.

POLCH, E.Z., CRUSE, T.A. & HUANG, C.-J. (1987). Traction BIE solutions for flat cracks, *Comput. Mech.*, **2**, 253–67.

RICKER, N. (1945). The computation of output disturbances from amplifiers for true wavelet inputs, *Geophysics*, **10**, 207–20.

RIZZO, F.J. & SHIPPY, D.J. (1970). A method for stress determination in plane anisotropic elastic bodies, *J. Comp. Mater.*, **4**, 36–55.

SLÁDEK, V. & SLÁDEK, J. (1984). Transient elastodynamic three-dimensional problems in cracked bodies, *Appl. Math. Model.*, **8**, 2–10.

THAU, S.A. & LU, T.-H. (1971). Transient stress intensity factors for a finite crack in an elastic solid caused by a dilatational wave, *Int. J. Solids Struct.*, 7, 731–50.

VOGEL, S.M. & RIZZO, F.J. (1973). An integral equation formulation of three dimensional anisotropic elastostatic boundary value problems, *J. Elast.*, 3 203–16.

WILSON, R.B. & CRUSE, T.A. (1978). Efficient implementation of anisotropic three-dimensional boundary integral equation stress analysis, *Int. J. Numer Methods Eng.*, 12, 1383–97.

WONG, H.L. & LUCO, J.E. (1978). Dynamic response of rectangular foundations to obliquely incident seismic waves, *Earthquake Eng. Structural Dyn.*, 6, 3–16.

Chapter 6

BEM ANALYSIS OF PROBLEMS
OF FRACTURE MECHANICS

S.T. RAVEENDRA and T.A. CRUSE
Southwest Research Institute, San Antonio, Texas, USA

SUMMARY

This chapter describes the application of the boundary element method to three-dimensional fracture problems. Since the BEM solution equations for crack problems degenerate in the presence of coplanar crack surfaces, the multi-region approach is used to separate each crack surface into a different region. The numerical modelling accuracy is improved by representing the crack tip field accurately using quarter-point and traction singular elements at the crack front. A computer program based on this approach is used to evaluate the stress intensity factors for arbitrary cracks in three-dimensional bodies.

6.1 INTRODUCTION

In most instances, the residual fatigue life of a cracked structural element is characterised by the crack growth rate that may be correlated to the cyclic change in the crack tip stress intensity factor. The effects of the stress field, the crack size and shape, and the structural geometry are related to the stress intensity factor in a systematic manner. This parameter may be evaluated from stress analysis of the body. This chapter describes the evaluation of stress intensity factors for arbitrary cracked three-dimensional bodies using the boundary element method (BEM). The BEM is adopted since it is well known to be suitable for the analysis of rapidly varying stress fields associated with fracture mechanics. The purpose of this chapter is to illustrate some of the basic modelling problems associated with three-dimensional fracture mechanics analysis and to provide numerical results for complex cracked bodies.

Basic aspects of linear elastic fracture mechanics (LEFM) analysis and an overview of BEM modelling as well as the numerical computations associated with BEM calculations of fracture mechanics parameters are reviewed. Numerical results for three-dimensional cracked structures are given in the final section.

6.2 AN OVERVIEW OF LEFM AND BEM ANALYSES

6.2.1 Linear Elastic Fracture Mechanics Analysis

Material deficiencies in the form of pre-existing flaws initiate cracks and fractures in structures. The presence of a crack in a structural element generally induces high stress concentration at the crack tip and thereby reduces the strength of the structure. Fracture mechanics provide satisfactory means for the characterisation of these local crack tip stress fields as well as the elastic deformations of the material in the neighbourhood of the crack. In LEFM, the inelastic deformation in the vicinity of the crack tip due to stress concentrations is deemed to be small compared to the size of the crack and other characteristic lengths.

Elastic modelling of crack tip behaviour makes use of deformation due to three primary modes of loading, as illustrated in Fig. 1. The three modes are: the opening mode (Mode I) due to normal stress; the sliding mode (Mode II) due to in-plane shear stress; and the tearing mode (Mode III), due to out-of-plane shear.

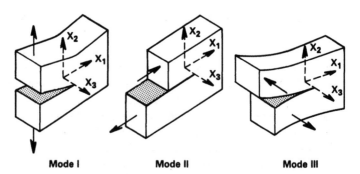

FIG. 1 Three primary loading modes of a cracked body.

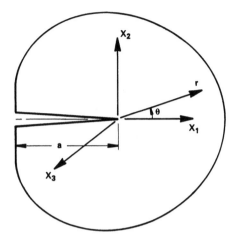

FIG. 2   Infinite plate under triaxial load.

Consider the crack problem of Fig. 2, representing an infinite plate under triaxial stress. The stresses and displacements for traction free cracks may be given as an infinite series in r, where r is the distance from the crack tip. For the anti-plane problem (see Kanninen and Popelar, 1985), the near crack tip field is given by

$$\begin{Bmatrix} \sigma_{31} \\ \sigma_{32} \end{Bmatrix} = \frac{K_{III}}{(2\pi r)^{1/2}} \begin{Bmatrix} \sin(\theta/2) \\ \cos(\theta/2) \end{Bmatrix} \tag{1}$$

and

$$u_3 = \frac{2K_{III}}{\mu} \left(\frac{r}{2\pi} \right)^{1/2} \sin(\theta/2) \tag{2}$$

where u_i is the displacement, σ_{ij} is the stress and μ is the shear modulus. The Mode III stress intensity factor, K_{III}, is established by the far field boundary conditions and is a function of applied loading and the geometry of the cracked body. The Mode III stress intensity factor is defined by

$$K_{III} = \lim_{r \to 0} \left\{ (2\pi r)^{1/2} \sigma_{32}|_{\theta=0} \right\} \tag{3}$$

For the plane problem, assuming plane strain conditions,

$$\begin{Bmatrix} \sigma_{11} \\ \sigma_{12} \\ \sigma_{22} \end{Bmatrix} = \frac{K_I}{(2\pi r)^{1/2}} \cos(\theta/2) \begin{Bmatrix} 1 - \sin(\theta/2)\ \sin(3\theta/2) \\ \sin(\theta/2)\ \cos(3\theta/2) \\ 1 + \sin(\theta/2)\ \sin(3\theta/2) \end{Bmatrix} \tag{4}$$

and

$$\begin{Bmatrix} u_1 \\ u_2 \end{Bmatrix} = \frac{K_I}{2\mu}\left(\frac{r}{2\pi}\right)^{1/2} \begin{Bmatrix} \cos(\theta/2)[k-1+2\sin^2(\theta/2)] \\ \sin(\theta/2)[k+1-2\cos^2(\theta/2)] \end{Bmatrix} \qquad (5)$$

where K_I is the Mode I stress intensity factor, defined by

$$K_I = \lim_{r\to 0}\left\{(2\pi r)^{1/2}\sigma_{22}|_{\theta=0}\right\} \qquad (6)$$

and

$$k = 3 - 4v$$

The corresponding relationships for Mode II fields are

$$\begin{Bmatrix} \sigma_{11} \\ \sigma_{12} \\ \sigma_{22} \end{Bmatrix} = \frac{K_{II}}{(2\pi r)^{1/2}} \begin{Bmatrix} -\sin(\theta/2)[2+\cos(\theta/2)\cos(3\theta/2)] \\ \cos(\theta/2)[1-\sin(\theta/2)\sin(3\theta/2)] \\ \sin(\theta/2)\cos(\theta/2)\cos(3\theta/2) \end{Bmatrix} \qquad (7)$$

$$\begin{Bmatrix} u_1 \\ u_2 \end{Bmatrix} = \frac{K_{II}}{2\mu}\left(\frac{r}{2\pi}\right)^{1/2} \begin{Bmatrix} \sin(\theta/2)[k+1+2\cos^2(\theta/2)] \\ -\cos(\theta/2)[k-1-2\sin^2(\theta/2)] \end{Bmatrix} \qquad (8)$$

where the Mode II stress intensity factor, K_{II}, is defined by

$$K_{II} = \lim_{r\to 0}\left\{(2\pi r)^{1/2}\sigma_{12}|_{\theta=0}\right\} \qquad (9)$$

6.2.2 Boundary Element Method for Fracture Mechanics Analysis

The analytical basis of the method is the transformation of the governing equilibrium equation of an isotropic, homogeneous, elastic element by an integral identity, using Betti's reciprocal work theorem. The identity for the displacement at a point $P(x)$ is given by (e.g. Banerjee and Butterfield, 1981):

$$C_{ij}(P)u_j(P) = -\int_S T_{ij}(P,Q)u_j(Q)\,ds(Q) + \int_S U_{ij}(P,Q)t_j(Q)\,ds(Q) \qquad (10)$$

where $t_i(Q)$ and $u_i(Q)$ are the boundary values of traction and displacement, and $T_{ij}(P,Q)$ and $U_{ij}(P,Q)$ are tractions and displacements, respectively, in the x_i directions at $Q(x)$ due to orthogonal unit loads in the x_j directions at $P(x)$. The discontinuity term C_{ij} is equal to 1/2 for smooth boundary points and can be evaluated indirectly using rigid body translation as described by Cruse (1975), for non-smooth boundary points.

The utility of the method as a general practical solution tool is facilitated by two approximations: one is the description of the boundary S by a finite number of surface elements; the second is the representation of the field variables (u_i, t_i) and geometry by known interpolation functions within individual elements. In the present analysis, triangular and quadrilateral elements are used for surface representation. The field variables, as well as geometry, are represented by isoparametric quadratic interpolation functions. Numerical evaluation of discretised integrals, as well as the use of special crack tip elements employed in the current analysis, are described in a later section.

6.2.3 BEM Modelling of Cracked Bodies
The numerical solution of eqn (10) is straightforward for a general three-dimensional stress analysis. However, the presence of two co-planar surfaces precludes the use of the method for general solution of cracked bodies. Therefore, several different modelling strategies have been employed for three-dimensional cracked bodies, as illustrated in Fig. 3. The

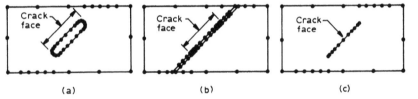

FIG. 3   BEM modelling strategies for three-dimensional fracture mechanics analysis: (a) open notch modelling; (b) multi-region modelling; (c) dislocation or traction BIE modelling.

first approach is to model the crack as an open notch as reported by Cruse (1972). The major deficiency of this modelling approach is that the results become inaccurate when the surfaces are modelled too far apart; however, the system equation becomes badly conditioned when the surfaces are modelled too close together. One form of avoiding this difficulty is the dislocation or traction BIE modelling approach, as developed in different forms by Cruse (1975), Guidera and Lardner (1975), Bui (1977) and Weaver (1977). The singular nature of the integrals in this method poses difficulty in the numerical implementation as reported by these authors. Significant improvements have been reported recently by Polch et al. (1987) and others. However, further research is required before the full potential of this utility can be realised.

Modelling of symmetric cracks is rather straightforward since only a symmetric part of the body that contains one crack surface needs to be modelled. Earlier application of this modelling has been reported by Cruse and VanBuren (1971), Cruse (1975), and Cruse and Meyers (1977). These results have been subsequently improved by using isoparametric interpolation functions by Cruse and Wilson (1977). The accuracy of the results has been further improved by the use of special crack tip elements that accurately model the crack tip field. Improved results using higher-order interpolation functions for geometry and field variables and special crack tip elements are also reported by Cruse and Wilson (1977). A detailed description of special crack tip elements is given in the following section.

A modelling strategy applicable for a general three-dimensional non-symmetric crack is the sub-region model. In this approach, the body is substructured into separate regions through the crack plane such that each crack surface is in a different region. The overall solution is obtained by satisfying compatibility and continuity conditions at the interface of the regions except along the crack surface. Numerical results using this strategy for two-dimensional structures have been reported by Blandford *et al.* (1981). Results for three-dimensional bodies are reported in this chapter.

6.2.4 Use of Singular Elements

The accuracy of the numerical computations is enhanced by proper representation of the field variables in the vicinity of the crack tip. It is well known that the crack tip opening displacement varies with the square of the distance (r) from the crack tip, whereas the stresses produce $1/\sqrt{r}$ singularity. The crack tip elements are modified such that they capture these variations.

It has been observed by Barsoum (1976) that the placement of midpoint nodes of the element sides emanating from the crack tip at quarter points leads to the required displacement and stress variation in the finite element method. By using quarter point (QP) elements (Fig. 4) on both sides of the

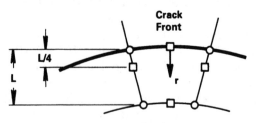

FIG. 4.   Crack-tip elements.

crack in BEM, the displacements and tractions are made to vary in physical space as:

$$\left\{ \begin{matrix} u(r) \\ \hat{t}(r) \end{matrix} \right\} = A_1 + A_2 \sqrt{(r/l)} + A_3(r/l) \tag{11}$$

where r is the normal distance from the crack tip, l is the element length in that direction, and A_1, A_2, A_3 are functions of nodal displacements or tractions. However, unlike the finite element method in which the tractions (or stresses) are derived from the spatial derivative of displacements, the displacements and tractions are independently approximated in BEM. Thus the use of quarter point elements on both sides of the crack front will give the same variation for both displacements and tractions and, therefore, eqn (11) does not give the required singularity for crack tip tractions. The simplest way to obtain this singularity appears to be through the use of singular shape functions. However, the numerical integration scheme employed in the current analysis requires subsegmentation of elements, as described in the following section, which makes the implementation difficult. Instead, the traction singularity is contrived by the multiplication of the shape function by a non-dimensional parameter $\sqrt{(l/r)}$ as:

$$t(r) = \sqrt{(l/r)} \cdot \hat{t} \tag{12}$$

where $\hat{t}$ is the nominal traction defined by eqn (11). The variation of traction in these traction singular (TS) elements is then given by:

$$t(r) = \frac{A_1}{\sqrt{(r/l)}} + A_2 + A_3(r/l)^{1/2} \tag{13}$$

The use of these special elements improves the accuracy substantially as seen by the numerical results reported herein.

6.2.5 Stress Intensity Factor Evaluations

The stress intensity factors may be computed using displacements or traction BEM solution from eqns (1)–(9). Mode I and Mode II stress intensity factors are computed from

$$K_{\text{I or II}} = \frac{\mu}{4(1-\nu)} \sqrt{\left(\frac{2\pi}{r}\right)} u_{\text{I or II}}(r) \tag{14}$$

where u_{I} and u_{II} are decoupled displacements along Mode I and Mode II directions at the quarter point, and r is the distance of the quarter point from the crack front. Mode III stress intensity factors may be computed similarly from

$$K_{III} = \frac{\mu}{4} \sqrt{\left(\frac{2\pi}{r}\right)} u_{III}(r) \tag{15}$$

Alternatively, the stress intensity factors can be computed from decoupled tractions at the crack front. As an example, the Mode III stress intensity factor may be evaluated from

$$K_{III} = \sqrt{(2\pi l)} \hat{t}_{III} \tag{16}$$

where $\hat{t}_{III}$ is the decoupled nominal traction defined by eqn (12). However, in the present analysis, the stress intensity factors are computed using quarter point displacements (eqns (14) and (15)), since the values using crack front tractions (eqn (16)) generally overestimated the stress intensity factors by about 10%. It is believed that one of the reasons for the discrepancy may be that higher accuracy is needed for the integration of traction singular elements than the one used in the present analysis. Another reason may be that, though the first term in eqn (13) provides the required singularity, the higher-order terms of the standard quadratic element expression (11), through traction singular modification, are the suitable ones for the representation of traction variation. However, the use of traction singular modification with quarter point elements improved the accuracy of the displacement-based stress intensity factors, and these elements are therefore used in the current analysis.

6.2.6 Evaluation of Discretised Boundary Integrals

In the present analysis, the geometry as well as field variables are represented by isoparametric quadratic shape functions. The use of higher-order interpolation functions in general precludes the use of analytical integration, and therefore numerical quadrature is used in the current development. Non-singular kernel function-shape function products, in principle, can be directly approximated by the application of the Gauss–Legendre quadrature formula. However, to maintain a certain level of accuracy, element subdivision may be required. Following Lachat and Watson (1976), the minimum element side length for a given error tolerance and quadrature order is determined from error analysis. The element is then subdivided to satisfy this requirement.

The non-singular integration is performed through a polar coordinate transformation which eliminates the singularity. To accomplish this, the element is subdivided through the singular point and the polar coordinate system is constructed through the singular apex as described by Banerjee and Raveendra (1986).

To achieve the required crack tip field variation, considerable mesh refinement at the crack tip is necessary. Because of this non-gradual mesh refinement, the subdivision scheme employed for non-singular integration does not always work efficiently. In order to avoid a large number of subdivisions, a modified singular integration scheme is used for these non-singular integrations.

6.3 NUMERICAL SOLUTIONS

The solution procedure outlined in the previous section has been applied to the solution of many symmetric and non-symmetric cracked specimens. The numerical solutions obtained by using the current computer code CRX3D are validated against known analytical solutions in some cases, and are compared with finite element solutions in other cases.

6.3.1 Circular Crack

Figure 5 shows the BEM map used for the analysis of a buried circular crack problem. Due to symmetry, only one-eighth of the body was

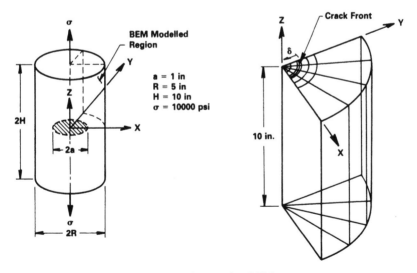

FIG. 5   Circular crack—BEM map.

modelled. In addition, the symmetric planes of the body were not modelled. The crack opening displacements using standard quadratic element and modified crack tip elements are compared to analytical results in Fig. 6 which indicates that the results are improved by the used of modified crack tip elements. The analytical solution plotted in the figure is for an infinite body. To further assess the accuracy, the Mode I stress intensity factors were evaluated for standard and modified crack-tip element models. A comparison with the empirical value that takes the finite dimension of the body into account demonstrated that the stress intensity factors using quarter point elements (QP) are approximately 5% in error, compared to standard element results which are approximately 10% in error. Further enhancement was obtained by using traction singular (TS) modification to the quarter point element, which improved the stress intensity factor value to within 1% of the predicted value.

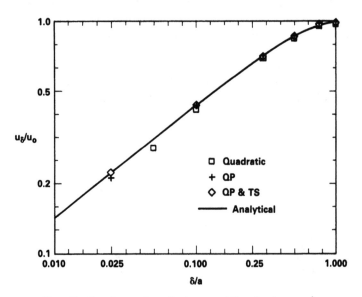

FIG. 6   Crack opening displacement for circular crack.

6.3.2 Elliptical Surface Crack

Figure 7 shows the BEM map for one-fourth of a finite cracked plate. Whereas the symmetric $Y-Z$ plane was not modelled, the $X-Z$ plane was modelled to allow the solution of elliptic buried and semi-elliptic surface cracks to be modelled by a change in boundary condition. Figure 8 shows

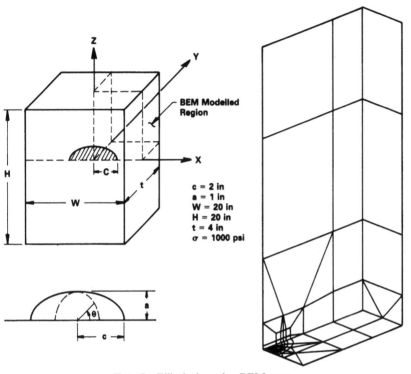

c = 2 in
a = 1 in
W = 20 in
H = 20 in
t = 4 in
σ = 1000 psi

FIG. 7   Elliptical crack—BEM map.

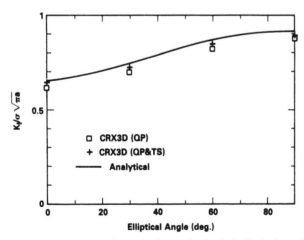

FIG. 8   Mode I stress intensity factor for buried elliptical crack.

the normalised Mode I stress intensity factor for the buried crack using
modified crack tip elements compared to the infinite body solution. The
figure indicates that the accuracy of the solution was again improved by the
use of the traction singular model.

The semi-elliptic surface crack model was subjected to both tensile and
bending loads. Figures 9 and 10 show good agreement between CRX3D

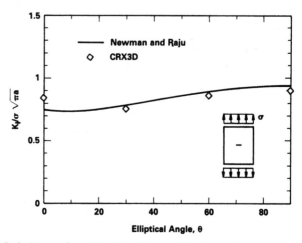

FIG. 9   Mode I stress intensity factor for semi-elliptic surface crack—tensile load.

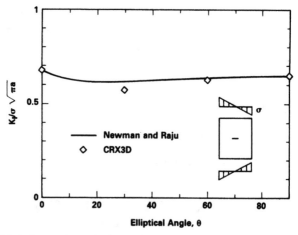

FIG. 10   Mode I stress intensity factor for semi-elliptic surface crack—bending load.

results and the empirical Mode I stress intensity factors provided by
Newman and Raju (1981) for both loading cases.

6.3.3 Inclined Circular Crack

Due to the symmetry, only one surface of the crack was modelled in the
previous examples. However, as explained in the previous section, the
non-symmetric cracks are modelled using subregion modelling strategy.
CRX3D code was validated for non-symmetric crack cases by analysing
a circular crack which is inclined 45° to the loading direction. Figure 11

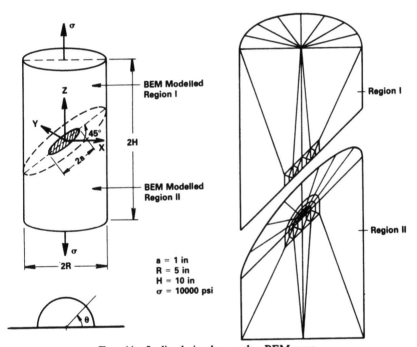

FIG. 11   Inclined circular crack—BEM map.

shows the BEM model of one-half of the cracked body. Stress intensity
factors for all three modes compared well with the infinite body analytical
solution, as shown in Fig. 12.

6.3.4 T-joint with Elliptical Surface Flaw

The final example solved is a T-joint section (Fig. 13(a)) that comprises
a part-elliptic surface flaw. A BEM model of one-half of the body is shown

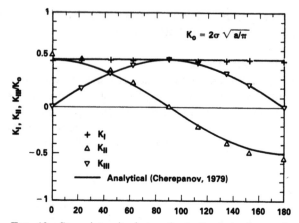

FIG. 12   Stress intensity factors for inclined circular crack.

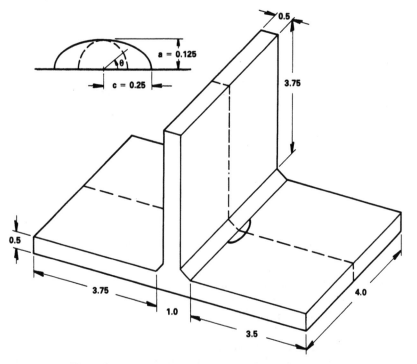

FIG. 13(a)   T-Joint section with elliptical surface flaw.

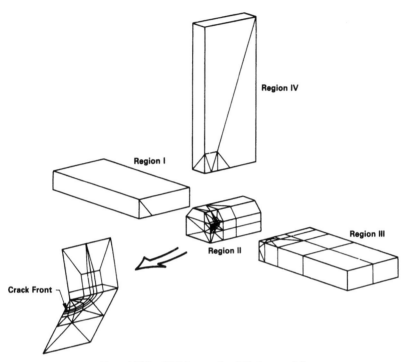

FIG. 13(b)   BEM map for T-joint model.

in Fig. 13(b). Again, a substructure modelling strategy was used. The body was modelled into four subregions to further improve the accuracy. Figure 15 shows the Mode I stress intensity factor normalised with respect to a two-dimensional plane strain model result (Fig. 14). The behaviour indicates that while the stress intensity factor is higher at the surface of the body, the solution at the mid-surface is comparable to the two-dimensional solution as expected. Mode II and Mode III stress intensity factors are also computed for the model and are shown in Fig. 16.

6.4 CONCLUSION

A three-dimensional boundary element code (CRX3D) was developed for the solution of arbitrary cracked bodies. In addition to cracks which are symmetric with respect to the geometry and loading, non-symmetric cracks

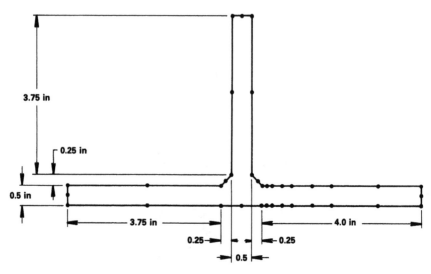

FIG. 14   BEM map for plane strain T-joint model.

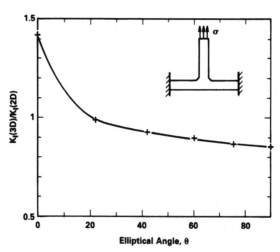

FIG. 15   Mode I stress intensity factor for T-joint.

are solved by using substructure modelling technique. To model near crack
tip displacement and traction fields accurately, modified crack tip elements
are used. The numerical results show that accurate solutions for complex
geometry and boundary conditions can be obtained by this procedure.

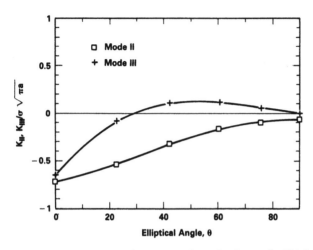

FIG. 16 Mode II and Mode III stress intensity factors for T-joint.

ACKNOWLEDGEMENTS

The authors wish to thank the Air Force Office of Scientific Research for partial support of this work through contract F49620-86-C-0048. Appreciation is extended to Mr Stan Dobosz for preparing the BEM models and to Ms Tanya Jackson for her assistance in preparing the manuscript. The first author wishes to express sincere thanks to Professor P.K. Banerjee under whose guidance part of the computer code was developed.

This manuscript is submitted for publication with the understanding that the United States Government is authorised to reproduce and distribute reprints for governmental purposes notwithstanding any copyright notation hereon.

REFERENCES

BANERJEE, P.K. & BUTTERFIELD, R. (1981). *Boundary Element Methods in Engineering Science*, McGraw-Hill, London. Also US Edition, New York, 1983; Russian Edition, Moscow, 1984.

BANERJEE, P.K. & RAVEENDRA, S.T. (1986). Advanced boundary element analysis of two- and three-dimensional problems of elastoplasticity, *Int. J. Numer. Methods Eng.*, **23**, 985–1002.

BARSOUM, R.S. (1976). On the use of isoparametric finite elements in linear fracture mechanics, *Int. J. Numer. Methods Eng.*, **10**, 25–37.

204   S.T. RAVEENDRA AND T.A. CRUSE

BLANDFORD, G.E., INGRAFFEA, A.R. & LIGGETT, J.A. (1981). Two-dimensional stress intensity factor computations using the boundary element method, *Int. J. Numer. Methods Eng.*, **17**, 387–404.
BUI, H.D. (1977). An integral equation method for solving the problem of a plane crack of arbitrary shape, *J. Mech. Phys. Sol.*, **25**, 29–37.
CHEREPANOV, G.P. (1979). *Mechanics of Brittle Fracture*, McGraw-Hill, New York.
CRUSE, T.A. (1972). Numerical evaluation of elastic stress intensity factors by the boundary-integral equation method, in: *The Surface Crack: Physical Problems and Computational Solutions*, American Society of Mechanical Engineers, 153–70.
CRUSE, T.A. (1975). *Boundary-integral equation method for three-dimensional elastic fracture mechanics analysis*, AFOSR-TR-75-0813, Accession No. ADA 011660.
CRUSE, T.A. & MEYERS, G.J. (1977). Three-dimensional fracture mechanics analysis, *J. Struct. Div. Am. Soc. Civ. Eng.*, **103**, 309–20.
CRUSE, T.A. & VANBUREN, W. (1971). Three dimensional elastic stress analysis of a fracture specimen with an edge crack, *Int. J. Fract. Mech.*, **7**, 1–15.
CRUSE, T.A. & WILSON, R.B. (1977). *Boundary-integral equation method for elastic fracture mechanics analysis*, AFOSR-TR-0355, Accession No. ADA 051992.
GUIDERA, T.J. & LARDNER, R.W. (1975). Penny-shaped cracks, *J. Elast.*, **5**, 59–73.
KANNINEN, M.F. & POPELAR, C.H. (1985). *Advanced Fracture Mechanics*, Oxford Science Series 15, Oxford University Press, New York.
LACHAT, J.C. & WATSON, J.O. (1976). Effective numerical treatment of boundary integral equations: a formulation for three-dimensional elastostatics, *Int. J. Numer. Methods Eng.*, **10**, 991–1005.
NEWMAN, J.C. & RAJU, I.S. (1981). An empirical stress-intensity factor equation for the surface crack, *Eng. Fract. Mech.*, **15**, 185–92.
POLCH, E.Z., CRUSE, T.A. & HUANG, C.-J. (1987). Traction BIE solutions for flat cracks, *Comput. Mech.*, **2**, 253–67.
WEAVER, J. (1977). Three dimensional crack analysis, *Int. J. Solids Struct.*, **13**, 321–30.

Chapter 7

BOUNDARY ELEMENT APPLICATIONS
IN THE AUTOMOTIVE INDUSTRY

W.E. FALBY and P. SURULINARAYANASAMI

Ford Motor Company, Dearborn, Michigan, USA

SUMMARY

This chapter summarises the use of BEM in the US automotive industry, especially at the Ford Motor Company. In spite of its enormous potential, BEM has not been extensively used in the automotive industry in the USA, largely due to the lack of availability of reliable engineering software. This chapter describes the experiences of using some of the available software in the analyses of selected automotive components.

7.1 INTRODUCTION

The solution of continuum mechanics problems, such as solid mechanics, acoustics, and heat transfer, has been made easier in the last 30 years, because of the development and computer application of two powerful numerical analysis methods. These are called the finite difference method, and the finite element method. In each of these methods, the strategy is to convert the governing differential equation within the domain where it applies into a system of simultaneous algebraic equations. These equations are solved by computer routines that are well known.

In the first method (the finite difference method), the domain is modelled by a set of grid points. The governing differential equation is then approximated by difference equations which, when applied to patterns of grid points, systematically produce the system of simultaneous equations required. The application of this method becomes very cumbersome and difficult for three-dimensional domains of irregular shape. A recourse to crude shape-approximations causes inaccurate results.

In the second method, called the finite element method, the total domain is divided into a set of regularly shaped adjoining regions such as bricks and tetrahedra. These regions are called finite elements. A generalised solution is then developed for each finite element. The elements are tied together by enforcing the continuity and boundary conditions of the system. The simultaneous equations obtained are solved using some excellent solution schemes developed specifically for these applications.

On the one hand, since the late 1960s, the finite element method has revolutionised analysis in the automotive industry by its power and range of applicability. On the other hand, solutions to larger and more complex problems are now demanded, together with the necessary computer resources. Thus, existing computer resources soon become overtaxed. Additionally, the considerable time required for the preparation of input data, and for interpreting output data, precludes the use of the finite element method in the redesign situations so characteristic of conceptual design, at least in its present form.

The two methods of numerical analysis described above are known as domain methods, because they directly discretise the domain where the governing differential equation applies. In the boundary element method only the surface enveloping the three-dimensional domain of interest, not the domain itself, is discretised with a system of space quadrilaterals and triangular surfaces called boundary elements. The main advantage of this method is that the boundary element model is a two-dimensional model, not a three-dimensional one as employed in the domain methods. The resulting system of equations is much smaller than that obtained from finite element modelling. Additionally, the cost and time required for preparing input data and interpreting output data are much reduced.

In spite of the advantages presented by the use of the boundary element method, the method is not gaining acceptability at the rate forecast for it. Most engineers feel that they already meet all deadlines with the finite element method and that there is no urgency to change. Recent surveys conducted at the Ford Motor Company have disclosed that engineers do not understand the method. In particular, modelling a domain gives them a physical intuition that is absent when a surface is modelled instead. In other words, they do not understand the volume to surface conversion. For this and other reasons, they do not believe that the boundary element method can deliver the advantages claimed for it.

In the next section, the basic mathematical principles of the boundary element method (BEM) will be set forth.

7.2 MATHEMATICAL THEORY OF BEM

The governing differential equation of any problem in engineering can be symbolically expressed as

$$L(\mathbf{u}) + \mathbf{b} = 0 \tag{1}$$

where L is a linear differential operator, $\mathbf{u}$ the solution variable and $\mathbf{b}$ a known linear quantity or an unknown non-linear term which depends on the solution $\mathbf{u}$ or its derivatives.

If we introduce a new solution $\mathbf{u}^*$ which is assumed to be continuously differentiable as many times as required, multiply eqn (1) by it and integrate by parts as many times as the order of the differentiation in operator $L(\mathbf{u})$, we can obtain an integral identity as shown below (Banerjee and Butterfield, 1981):

$$\int_V [L(\mathbf{u}) + \mathbf{b}]\mathbf{u}^* \, dv = 0$$

or

$$\int_S [M(\mathbf{u})\mathbf{u}^* - M(\mathbf{u}^*)\mathbf{u}] \, ds + \int_V \mathbf{b}\mathbf{u}^* \, dv = \int_V L^\circ(\mathbf{u}^*)\mathbf{u} \, dv \tag{2}$$

where
M is a differential operator on the S,
L° is a new differential operator which for classical governing equations often turns out to be an adjoint operator of L.

If we now choose $\mathbf{u}^*$ to be the solution of the equation

$$L^\circ(\mathbf{u}^*) + \delta(x, \xi) = 0 \tag{3}$$

where $\delta(x, \xi)$ is a unit impulse function at ξ, we can easily see that the right-hand side of eqn (2) can be written as

$$\int_V L^\circ(\mathbf{u}^*)\mathbf{u} \, dv = \int_V \delta(x, \xi)\mathbf{u}(x) \, dv = \mathbf{u}(\xi) \tag{4}$$

Solution u^* which satisfies eqn (3) is called the 'fundamental solution'. Often these are point force or point source solutions in an infinite solid.

Substituting eqn (4) in eqn (2) we can obtain

$$u(\xi) = \int_S [M(u)u^* - M(u^*)u] \, ds + \int_V bu^* \, dv \qquad (5)$$

Equation (5) is now a general integral statement of the boundary value problem. Usually the boundary element solution is then developed by taking the point ξ on the boundary nodes of the discretised problem (by dividing the boundary into surface elements and the volume into several volume cells), leading to the matrix equations (Banerjee and Butterfield, 1981):

$$Gt - Fu + Cb = 0 \qquad (6)$$

In a given boundary value problem the boundary quantities t and u are given in such a manner that exactly half are prescribed and the remaining half need to be determined. Equation (6) can then be rearranged to form (Banerjee and Butterfield, 1981):

$$Ax = By - Cb \qquad (7)$$

where x and y are known and unknown quantities over the boundaries.

Equation (7) can now be solved for x if b is prescribed or can be determined iteratively for a non-linear problem. For most linear problems the volume integral in eqn (5) can be converted into a surface integral using the divergence theorem and, as a consequence, the problem becomes one of boundary discretisation only. Alternatively it is possible to use the method of particular integrals to account for these volume effects (Banerjee et al., 1988).

The solution of eqn (1) can also be constructed as the sum of the complementary function u^c and a particular integral u^p where these satisfy the equations

$$L(u^c) = 0 \qquad (8)$$

$$L(u^p) + b = 0 \qquad (9)$$

The particular solution u^p is any solution (algebraic polynomial) which satisfies eqn (9).

The solution of eqn (8) by BEM will now lead via eqn (5) with $b = 0$ to a boundary only system

$$Gt^c - Fu^c = 0 \qquad (10)$$

which can be modified by the particular solution u^p and its derivative t^p as

$$Gt - Fu = Gt^p - Fu^p \qquad (11)$$

where the total solutions u and t are taken as the sum of the complementary solutions u^c and t^c and particular integrals u^p and t^p. Equation (11) can now be reorganised in terms of known and unknown quantities as

$$Ax = By + Gt^p - Fu^p \qquad (12)$$

in which the right-hand side is either known in a linear problem or can be determined from the information pertaining to it in a non-linear problem.

The two mathematical strategies outlined above have been developed by Banerjee and his co-workers for a very large class of problems in engineering. These range from linear and non-linear problems of stress anaysis to dynamic stress anaysis, acoustics, fluid flow, heat transfer, thermoviscous fluid flow, etc. These developments inspired the authors to re-examine BEM once again for automotive industrial applications.

7.3 BOUNDARY ELEMENT EXPERIENCE IN THE AUTOMOTIVE INDUSTRY

The BEM was introduced to the domestic automotive industry at the Ford Motor Company by Bozek et al. (1979). At that time, the company was attempting to solve large stress analysis problems, but was being restricted by the limited computational capacity available. It was thought that this problem could be solved with the aid of the boundary element method. The project called for first analysing a coarse finite element method (FEM) model of the structure, then for analysing a more refined BEM model of the localised areas of interest, using boundary conditions obtained from the previous FEM results. The FEM code used was MSC NASTRAN (1965). The BEM code used was described by Rizzo and Shippy (1977); this was not a user-friendly commercial code.

It was determined that the approach would be very effective for the analysis of complex three-dimensional shapes. However, the application of BEM directly to an appropriate model rather than to a BEM model in tandem with the FEM model was even more effective. For analyses conducted on BEM and FEM models with identical surface grids, the BEM produced substantially more accurate results than the FEM. On the basis of computer running times, however, there was no clear advantage in using the BEM over FEM. It should be noted, however, that BEM has undergone about 10 years of development since then.

There was little follow-up on this BEM initiative for several years. In Bozek (1983) parallel processing was employed to reduce the run times necessary for some BEM analyses. The component analysed was a drive shaft trunnion. Although results were encouraging, further work needs to be done in this area before this method could be applied to general problems.

In recent years, BEM has undergone substantial development and it has become apparent that it can compete, in its own right, with the finite element method, were it not for the degree to which the latter is entrenched in the user community. Additionally, it was determined that in order to reduce the duration of the vehicle design cycle (now about six to seven years), and to 'get it right the first time', it would be necessary to introduce computer-aided engineering in the preliminary design and redesign stages. Boundary element models were more adaptable to updating due to design modifications, than are finite element models. This issue is discussed further in Section 7.4 below. Motivated by these problems, it was decided to give the boundary element method a second look, particularly since at this time also several commercial-type BEM codes began to appear on the market.

In mid-1985, the code BEASY (Boundary Element Analysis System), described in Danson *et al.* (1982), was loaned to the Ford Motor Company for evaluation by Computational Mechanics, Inc. of Woburn, Mass. The code could only be exercised in its small configuration mode. Thus, it was possible to solve only small problems with it.

Also towards the end of 1985, Ford acquired a NASA-funded boundary element code for evaluation. That code was called BEST3D (Boundary Element Solution Technology) by Banerjee and Wilson (1986). Although the capabilities of this code extended from stress analysis to recently developed algorithms for heat transfer, it was intended mainly for stress analysis applications. It was soon realised that opportunities for technology transfer existed in several areas. One of these areas was vehicle road boom, the focus of the next section. Initially, this capability was developed specifically for the Ford Motor Company. The code has been recently upgraded and rewritten. Its new name is GPBEST (General Purpose BEST) (Banerjee, 1987).

A third boundary element code, EZBEA (Easy Boundary Element Analysis), an evolution of the early code of Rizzo and Shippy (1977) referred to above, was recently introduced to the Ford Motor Company (EZBEA, 1987).

Each of these codes has different strengths and weaknesses. For instance, Manolis and Banerjee (1986) showed that both BEST3D and GPBEST were strong in problem formulation and numerical implementation. These codes use conforming boundary elements, resulting in superior running times, excellent convergence characteristics, and high accuracy of results. On the other hand, the codes BEASY and EZBEA are thought to be user-friendly. However, all these codes remain in a state of modification and growth. The reader is, therefore, advised to get in touch with the codes' developers to ascertain their present status. For convenience, the recent capabilities of each of these codes are summarised below.

For the sake of comparison, the capabilities of the BEM computer codes used at Ford will be described briefly. By capabilities is meant the types of analysis the code could do. It also means features that influence the codes' performance, positively or negatively. These descriptions are valid for the period of use only. For instance, it is understood that BEASY has undergone substantial changes in recent months, including the introduction of partially conforming elements. On the other hand, earlier versions of BEST3D did not contain a modal analysis capability, but it now does.

The capabilities of the code BEASY were as follows:

— Static elastic analysis
— Steady state heat transfer
— Electrical conduction
— Time-dependent potential analysis

The code had excellent pre-processing and post-processing tools. The computer execution times of this code for stress analysis were surprisingly long.

The code GPBEST had the following capabilities:

— Static elastic analysis
— Steady state and transient thermoelastic stress analysis
— Static elastoplastic analysis
— Free and force vibration
— Heat transfer (steady state and transient)
— Thermal and body force load via particular integral
— Transient elastodynamics including non-linear stress analysis
— Geotechnical analysis
— Fluid flow
— Acoustics (periodic, transient and eigenfrequency)

The pre- and post-processing capabilities of this code are linked to PATRAN.

A surprising feature of the code was that for acoustic analysis of comparable passenger compartment models, the modal extraction was twice as efficient, with regard to computer execution time, as that of the MSC NASTRAN.

Finally, the code EZBEA had three capabilities:

— Static elastic analysis
— Steady state thermoelasticity
— Heat conduction analysis

It is clear from the above that BEM has really reached a very high level of development and maturity, and it is also apparent that such a powerful engineering analysis tool must be used in industrial applications where one analysis tool is usually not enough for complex problems. Engineers need to feel comfortable with the correctness of the results which cannot usually be established from one analysis. In the following section BEM applications in acoustic eigenfrequency analysis, stress analysis and vibration analysis are described to illustrate this point.

7.4 CONCEPTUAL DESIGN FOR ROAD BOOM

7.4.1 Introduction
The purpose of the conceptual design is to improve the quality of design for 'boom' in the vehicle at the conceptual and preliminary design stage. This is made possible by managing existing powerful CAE tools in a new way so that redesign turnaround time is substantially reduced. The trade-off would seem to be loss of accuracy. However, there is less here than meets the eye, since the initial design data are approximate at this stage anyway. As the quality of input data is improved, the CAE method is turned accordingly. Additionally, there is a departure from the conventional approach to modelling. For instance, where three-dimensional models were used in the past, two-dimensional models can now be employed to obtain similar accuracy. When the accuracy may be reduced, as in conceptual design situations, simple one-dimensional models may also be employed.

The booming noise in vehicle compartments, due to powertrain, road disturbance and other excitation, is called 'boom'. This is a fluid/structure interaction problem, and occurs in the 20 Hz to 60 Hz frequency range. Boom has become a more critical design problem, in recent years, because the automobile body is now less isolated from the excitation than when it was mounted on a full frame, and because of other construction changes that have been instituted to save weight.

According to Osawa and Iwama (1986), the following are the chief factors contributing to boom:

— Road roughness characteristics
— Vibration characteristics of tyres and suspensions
— Vibration characteristics of vehicle body
— Acoustic characteristic of the air in the passenger compartment, sometimes called the acoustic cavity

Since, as has been stated above, the frequency range of acute auditory pressure sensation at the driver's ear is known, one method of designing for boom is to ensure that the peak frequency of the booming noise lies outside the above range.

The present analytical procedure for boom (Nefske *et al.*, 1982; Flanigan and Borders, 1983) consists first of determining the frequencies (or eigenvalues) and mode shapes (or eignevectors) of the body structure and the acoustic cavity, separately, by the finite element method (FEM). The total system, including body, acoustic cavity, chassis and tyres, is then coupled and solved for the specified excitation. A block diagram of the procedure is given in Fig. 1. All the operations shown in the diagram can be

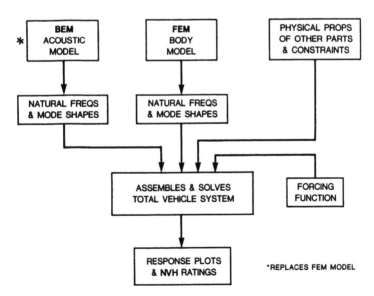

FIG. 1   Block diagram for boom determination (BEM acoustics).

performed by a FEM code like NASTRAN. However, it has been found more efficient to use special proprietary coding for the coupling and solution procedures.

A typical FEM model of an acoustic cavity of a vehicle is shown in Fig. 2. It is to be noted that, in this analysis, pressure levels are required at only two grid points within the cavity, one located at the driver's ear and the other located at the rear seat passenger's ear. However, in modelling for an FEM solution, it is necessary to describe the cavity with thousands of grid points located both within the cavity and on the structure fluid interface, with solid elements. The actual size of this model was 1031 solid elements and 1321 grid points. It is clear that such an extensive modelling of the entire region is not really practical for the numerous iterations that are necessary for conceptual design.

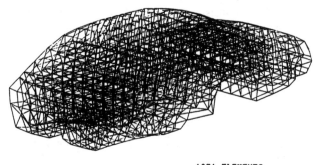

1031 ELEMENTS
1321 GRID POINTS

FIG. 2   FEM model of vehicle acoustic cavity.

An alternative formulation based on the boundary element method was recently developed by Banerjee *et al.* (1988) in the GPBEST program. This work was supported by the Ford Motor Company and is described below.

7.4.2 BEM Formulation for the Acoustic Eigenfrequency Analysis

The governing differential equation for the sound pressure $pe^{i\omega t}$ in a periodic acoustic problem can be written as

$$\frac{\partial^2 p}{\partial x_i \partial x_i} + q = -\left(\frac{\omega}{c}\right)^2 p \tag{13}$$

where

ω is the circular frequency in rad/s,
c is the speed of sound through the fluid,
$qe^{i\omega t}$ is the specified sound source inside the volume.

Equation (13) needs to be solved for the specified boundary conditions, i.e.

$p=0$ on the surface s_1 if the part of the surface is open,

$\dfrac{\partial p}{\partial n}=0$ on the surface s_2 if the part of the surface is hard.

Alternatively if the surface s_2 is vibrating, for a small amplitude motion, the fluid and surface interaction condition can be obtained from the momentum balance equation

$$\frac{\partial p}{\partial n} = -\rho\frac{\partial^2 u}{\partial t^2} \tag{14}$$

where ρ is the air density and u is the normal component of the displacement of the surface. For a periodic response of the surface displacement of the form $ue^{i\omega t}$ the above boundary condition can be written as

$$\frac{\partial p}{\partial n} = \rho\omega^2 u \tag{15}$$

For an eigenfrequency analysis, the source term q is taken as zero and the solution for the acoustic pressure p is taken as the sum of the complementary function p^c and a particular integral p^I. The complementary function p^c satisfies the homogeneous part of the differential equation

$$\frac{\partial^2 p^c}{\partial x_i \partial x_i} = 0 \tag{16}$$

and the particular integral p^I is any arbitrary polynomial-type expression satisfying the total differential equation

$$\frac{\partial^2 p^I}{\partial x_i \partial x_i} + \left(\frac{\omega}{c}\right)^2 p = 0 \tag{17}$$

Unfortunately p^I cannot be obtained from eqn (17) since the total solution p still remains in the total differential equation. In order to eliminate this difficulty, both Nardini and Brebbia (1982) and Ahmad and Banerjee (1986) used a global shape function to approximate this inhomogeneous term. For the present case one can use

$$p(x) = \sum_{m=1}^{\infty} C(x, \zeta^m)\phi(\zeta^m) \tag{18}$$

where

x is any point inside the region,
ζ^m is a boundary point at the node m,
ϕ is a fictitious function at the node m.

A simple function that is selected for the present analysis and found to work well is

$$C(x, \zeta^m) = (R - r)$$

where

R is the largest distance between any two points in the body, and
r is the distance between x and ζ^m.

With this approximation, the governing differential equation (17) becomes

$$\frac{\partial^2 p^l}{\partial x_i \partial x_i} + \left(\frac{\omega}{c}\right)^2 \sum_{m=1}^{\infty} (R - r)\phi(\zeta^m) = 0 \tag{19}$$

It can be easily seen that the following solution (Benerjee et al., 1988) will satisfy eqn (19):

$$p^l(x) = \sum_{m=1}^{\infty} D(x, \zeta^m)\phi(\zeta^m) \tag{20}$$

where

$$D(x, \omega^m) = \left(\frac{\omega}{c}\right)^2 [c_1 r^3 - c_2 R r^2]$$

$$C_1 = \frac{1}{3(d+1)}$$

$$C_2 = \frac{1}{2d}$$

$d = 2$ or 3, respectively, for two- or three-dimensional problems.

The normal derivative $\partial p^l/\partial n$ can then be written as

$$\frac{\partial p^l}{\partial n} = \sum_{m=1}^{\infty} T(x, \zeta^m)\phi(\zeta^m) \tag{21}$$

where

$$T(x, \zeta^m) = \left(\frac{\omega}{c}\right)^2 (3c_1 r - 2c_2 R)y_j n_j$$

$$y_j = x_j - \zeta_j^m$$

By using the usual surface discretisation and numerical integration, the boundary integral equation for the complementary part can be cast as

$$[G]\left\{\frac{\partial p^c}{\partial n}\right\} - [F]\{p^c\} = 0 \tag{22}$$

From eqns (20) and (21), with the nodes m identical to the boundary nodes, we can write

$$\{p^l\} = \lambda[D]\{\phi\}$$

$$\left\{\frac{\partial p^l}{\partial n}\right\} = \lambda[T]\{\phi\} \tag{23}$$

where

$$\lambda = \left(\frac{\omega}{c}\right)^2$$

These can now be inserted in eqn (22) to yield (Banerjee et al., 1988)

$$[G]\left\{\frac{\partial p}{\partial n}\right\} - [F]\{p\} = \lambda([G][T] - [F][D])\{\phi\} \tag{24}$$

But from eqn (18) one can write

$$\{p\} = [C]\{\phi\} \quad \text{or} \quad \{\phi\} = [C]^{-1}\{p\}$$

which can be used to eliminate $\{\phi\}$ in eqn (24) to form

$$[G]\left\{\frac{\partial p}{\partial n}\right\} - [F]\{p\} = \lambda[M]\{p\} \tag{25}$$

where

$$[M] = ([G][T] - [F][D])[C]^{-1}$$

By incorporating the appropriate boundary conditions, eqn (25) can be cast in a form for the eigenfrequency analysis as (Ahmad and Banerjee, 1986; Banerjee *et al.*, 1988):

$$[A]\{x\} - \lambda[B]\{x\} = 0 \qquad (26)$$

where the vector $\{x\}$ contains the unknown boundary values of p and $\partial p/\partial n$.

7.4.3 Application of GPBEST in Acoustic Eigenfrequency Analysis

Although GPBEST now has extensive external and internal acoustic analyses, such as periodic, transient and interior eigenfrequency analyses, only the eigenfrequency module which was developed for the Ford Motor Company will be discussed here.

As a necessary step to tackle much larger problems with this newly developed capability, a series of validation examples were undertaken by Banerjee *et al.* (1988). Some of these are reproduced below.

In order to investigate the accuracy of the proposed acoustic eigenfrequency analysis, a two-dimensional acoustic cavity of dimensions $b = 2$ m and $a = 40$ m was chosen. The speed of sound was taken as 340 m/s. Acoustically rigid (i.e. $\partial p/\partial n = 0$) boundary conditions were imposed on all surfaces. Since $a \gg b$, for lower modes, the behaviour would be similar to that of a one-dimensional problem for which the exact acoustic eigenfrequency (Hz) is given by

$$f = \frac{cn}{2a} = 4 \cdot 25n, \text{ where } n = 1, 2, 3 \ldots \left(\frac{a}{b}\right)$$

The convergence study for this problem is shown in Fig. 3 for the lowest five modes. For this analysis, the short side is always modelled by one quadratic element, while for the long sides variously 1, 2, 4, 6, 8, 10 and 12 elements were used. It can be seen that for one wavelength, the use of four elements in the longitudinal direction yields answers to within 2%. Even the fifth mode is well reproduced by taking eight elements in the longitudinal direction.

Shuku and Ishihara (1973) analysed the passenger compartment of a small car using a two-dimensional finite element model in which all the surfaces were assumed to be acoustic hard surfaces. They also carried out experimental studies on a model of the car with the seats removed. The same problem was analysed identically by using the present method and the results are compared in Table 1. The speed of sound was assumed to be 340 m/s. The mesh used in the boundary element analysis is shown in Fig. 4. It can be seen from the three reported natural frequencies in Table 1 that the experimental, finite element and boundary element results agree well. The

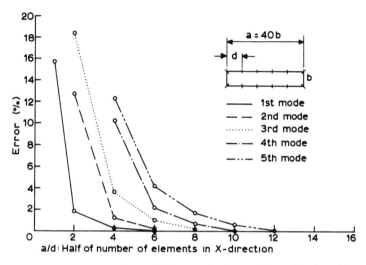

FIG. 3   Convergence of natural frequencies in one-dimensional problem.

TABLE 1

COMPARISON AMONG THE NATURAL FREQUENCIES (Hz) OF THE
TWO-DIMENSIONAL AUTOMOBILE COMPARTMENT (FIG. 2) BY
BEM, FEM AND EXPERIMENTAL METHODS

Mode	Experimental	FEM	BEM
1	87·5	86·8	87·6
2	138·5	138·0	138·7
3	157·0	154·6	153·2

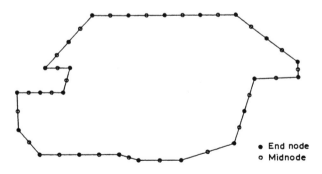

FIG. 4   BEM model of Toyota acoustic cavity.

maximum difference is 0·9% between the finite element and boundary element results, and about 2·5% between the boundary element results and the experimental results. The FEM model used 57 traingular elements and the BEM model shown in Fig. 4 has 23 quadratic elements.

In an excellent paper, Nefske *et al.* (1982) examined the acoustic eigenvalue problem for a truck cab using a three-dimensional finite element model. Unfortunately no dimensions of the model were given. The boundary element model shown in Fig. 5 has been generated for a cavity of approximately 7 ft × 6 ft × 5 ft 6 in overall dimensions, which was thought to be appropriate for a standard truck cab. The boundary element model uses 57 elements (two six-noded triangular elements and 55 eight-noded

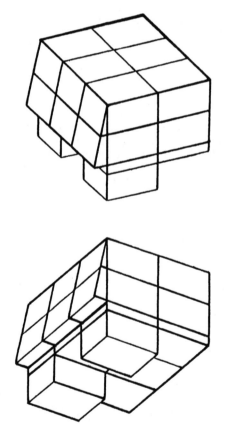

FIG. 5   Three-dimensional one-region boundary element model of a truck cab.

quadrilateral elements), leading to a 171-node problem. A 'rough' comparison between the results of Nefske *et al.* and the boundary element results are shown in Table 2. Once again the agreement is remarkable in view of the approximated geometry.

TABLE 2
COMPARISON BETWEEN THE NATURAL
FREQUENCIES (Hz) OF A TRUCK CAB
(FIG. 7) BY BEM AND FEM

Mode	Fem	Bem
1	67	72
2	82	85
3	105	109

These validation examples presented by Banerjee *et al.* (1988) did establish in the minds of the authors that BEM can be used for the acoustic boom analysis. Accordingly the vehicular cavity in Fig. 2 was analysed by a BEM model, shown in Fig. 6, which was generated from the finite element

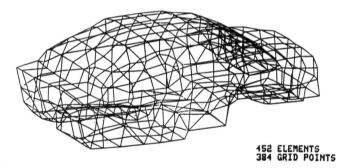

452 ELEMENTS
384 GRID POINTS

FIG. 6   BEM model of vehicle acoustic cavity.

data pertaining to the mesh shown in Fig. 2. A small computer program was developed to read the FEM data and pick up the necessary surface information. Additionally, a set of interior points near the driver's and passenger's ears was introduced. The BEM model shown uses 452 surface elements resulting in 384 nodes.

Typical computer times for the eigen/system extraction was about 1100 s for the FEM model and 500 s for the BEM model. These problems were run on a CRAY X-MP computer.

222   W.E. FALBY AND P. SURULINARAYANASAMI

In summary, the advantages claimed for the BEM in this application are as follows:

— The acoustic cavity problem is reduced from three dimensions to two dimensions, or from two dimensions to one dimension.
— Input data preparation and data recovery are reduced by an order of magnitude.

Because of these cost and time advantages, a designer will now have the ability to examine many more design options in preliminary design situations.

In order to demonstrate the usefulness of BEM in design situations, the vehicle compartment shown in Figs 2 and 6 was again examined. For simplicity, a two-dimensional longitudinal BEM model of the same acoustic cavity was prepared using 41 one-dimensional line elements, as shown in Fig. 7. The complexity introduced into both models was simply that the front and rear seats were modelled explicitly. The seats were represented in the manner discussed by Flanigan and Borders (1983). The seats were modelled as dense air with a density about 11 times that of the air in the rest of the cavity. The eigenvalues of the first two modes were chosen for comparison. The assumption, that a two-dimensional BEM acoustic model could predict the first two eigenvalues of a vehicle acoustic cavity, relied on the fact that the first two cavity modes are usually longitudinal modes.

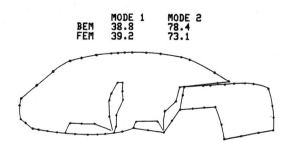

	MODE 1	MODE 2
BEM	38.8	78.4
FEM	39.2	73.1

FIG. 7   Frequencies (H_2) of vehicle acoustic cavity with seats.

The method, previously described, for predicting the pressure in the vehicle passenger compartment, for the boom, can be applied when the preliminary design is sufficiently advanced. However, when the design is at the conceptual level, it is controlled by 'bogeys'. For instance, we may be told to keep the first mode eigenvalue of the acoustic cavity below 38 Hz

MODE 1=93.7 HERTZ
MODE 2=171.6 HERTZ

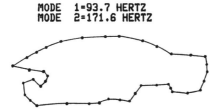

FIG. 8   BEM frequencies for conceptual design (one-dimensional model).

and above 47 Hz. In this case, a simplified model like that shown in Fig. 8 is ideal for enforcing this constraint. An optimisation method can be employed extremely well.

Another type of problem for which such a model is suitable, is demonstrated by the BEM models in Figs 8 and 9. Figure 8 represents an approximate BEM model of a new vehicle acoustic cavity. Figure 9 represents a modified BEM model where the modification consists of extending the base of the windshield 90 mm outward. It also shows the effect of the change on the first two modes. The first mode frequency is increased by 2·5 Hz, while the second mode is decreased by 2·9 Hz.

It can be easily concluded that BEM is an excellent tool for the acoustic analysis of an automobile compartment. Of course, it is well known that BEM does not really have any competition as far as exterior acoustic problems are concerned.

Original modes #1, 2: 93·7, 171·6
Modified modes #1, 2: 96·2, 168·7

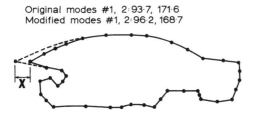

FIG. 9   Effect of acoustic cavity modification on frequencies. (X = 90 mm)

7.5 APPLICATIONS OF BEM IN ANALYSIS OF AUTOMOTIVE STRUCTURAL COMPONENTS

BEM has a long history of success in the area of solid mechanics. The current state of development is such that it can be almost routinely applied to both linear as well as non-linear problems. Two types of problems are being considered here, namely stress analysis and eigensystem analysis. These are the two major categories of solid mechanics analysis usually encountered in automotive work.

The first application presented here is the displacement analysis of the cross-section of a gear tooth from an automatic transmission. Thus the problem is a simple two-dimensional one as shown in Fig. 10. The tooth is fixed at its base, and a lateral load of 450 lb is applied at its tip as shown. The BEM model, Fig. 11, consists of 50 line-elements. The problem was analysed with BEASY, and was run on a CDC CYBER 176. The CPU time or run time was 6·5 s. The lateral displacement of the tooth tip obtained

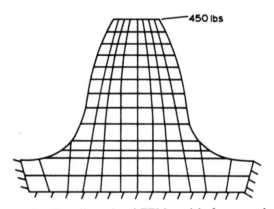

FIG. 10   Two-dimensional FEM model of gear tooth.

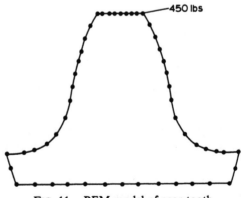

FIG. 11   BEM model of gear tooth.

from the NASTRAN analysis was 151×10^{-6} in compared to the BEASY result of 152×10^{-6} in.

The bearing cap from a tractor engine shown in Fig. 12 was analysed for a 0·0024 in lateral displacement at each end of its base. The displacement was imposed to force-fit the cap into a recess at the top of the engine block. Half of the cap had been analysed earlier with 53 three-dimensional solid brick elements by using NASTRAN. The GPBEST model, with 29 quadratic surface elements representing a quarter model of the bearing cap, is given in Fig. 13. A similar analysis of the cap was also made with BEASY.

The bending stresses were compared at the double symmetry points of the intrados and the extrados, for results from BEASY, GPBEST, NASTRAN, and test data, as shown in Table 3.

The GPBEST model was analysed initially using all quadratic elements; subsequently, a model with all linear elements was employed in a second analysis, with no appreciable change in the result. For the NASTRAN and GPBEST analyses a CRAY X-MP was used. The CPU times were 56·5 s for NASTRAN and 60·1 s for the GPBEST analysis using linear elements.

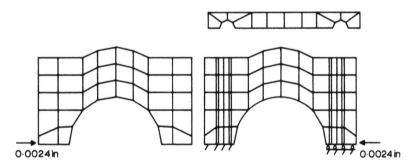

0·0024 in   0·0024 in

FIG. 12   FEM model of bearing cap.

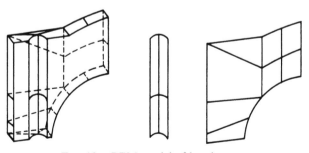

FIG. 13   BEM model of bearing cap.

TABLE 3

Location	Bending stress (psi)			
	Test	NASTRAN	BEASY	GPBEST
Top	3 584	3 440	3 221	2 816
Bottom	−9 100	−9 660	−7 844	−8 944

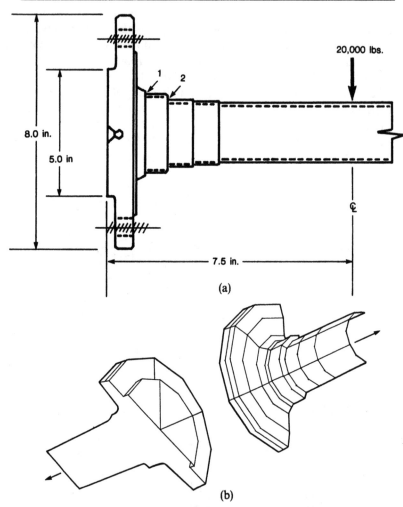

FIG. 14   (a) Section of loaded axle half-shaft (b) Rear axle half-shaft BEM model (quarter symmetry employed).

A rear axle half-shaft fixed as shown in Fig. 14(a), and subjected to a 20 000 lb load, was analysed by GPBEST for the principal stresses at locations 1 and 2. Only the top half of the shaft was modelled, and it contained 44 surface elements as shown in Fig. 14(b). The stresses obtained at locations 1 and 2 from GPBEST were 49 710 and 18 240 psi respectively, compared with those obtained from a previous NASTRAN analysis of 46 120 and 19 800 psi respectively. The computing times for both these analyses were once again very similar.

In order to explore the capabilities of GPBEST for the free vibration analysis, a modal analysis of the half-shaft problem discussed above was carried out using the mesh shown in Fig. 14(b). The results of the GPBEST analysis are compared with the NASTRAN results in Table 4. The first, second and fourth modes were almost identical. Unfortunately, NASTRAN was unable to obtain the third mode. A careful analysis of the problem did indicate that this was indeed a real mode. It is perhaps important to point out that for complex problems two independent analyses are often preferable, in order to establish if the results are indeed correct.

TABLE 4

Mode	NASTRAN $(Hz \times 10^4)$	GPBEST $(Hz \times 10^4)$
1	1·74	1·82
2	3·58	3·43
3	—	4·75
4	6·63	6·63

7.6 CONCLUSIONS

The implementation of boundary element methods in the automotive industry is still in its infancy. The types of problems at which BEM excels are now routinely analysed by MSC NASTRAN, an FEM code that has been in use for over a decade. However, it is anticipated that substantial inroads will be made in this area within the next few years. The use of BEM in acoustics will grow very quickly because of its clear advantages over FEM in such situations.

ACKNOWLEDGEMENTS

The authors are grateful to Professor P.K. Banerjee and Dr Gary Dargush of the State University of New York at Buffalo for helpful discussions and valuable suggestions.

REFERENCES

AHMAD, S & BANERJEE, P.K. (1986). A new method for vibration analysis by BEM using particular integrals, *J. Eng. Mech. Div., Am. Soc. Civ. Eng.*, 113, 682–95.

BANERJEE, P.K. (1987). *GPBEST User's Manual, Version 1.0*, 1987, Computational Mechanics Division, Department of Civil Engineering, State University of New York at Buffalo.

BANERJEE, P.K. & BUTTERFIELD, R. (1981). *Boundary Element Methods in Engineering Science*, McGraw-Hill, London. Also US Edition, New York, 1983; Russian Edition, Moscow, 1984.

BANERJEE, P.K. & WILSON, R.B. (1986). *BEST3D User's Manual*, Computational Mechanics Division, Department of Civil Engineering, State University of New York at Buffalo.

BANERJEE, P.K., AHMAD, S. & WANG, H.C. (1988). A new BEM formulation for acoustic eigenfrequency analysis, *Int. J. Numer. Methods Eng.*, 26, 1299–309.

BOZEK, D.G., KLINE, K.A. & MORMAN, K.N. (1979). Three-dimensional analysis using combined finite element–boundary integral techniques, *3rd Int. Conf. on Vechicle Structural Analysis*, Troy, Michigan.

BOZEK, D.G. *et al.* (1983). Vector processing applied to boundary element algorithms on the CDC CYBER—205, *EDF Bull. Dir. Etudes et Recherches, Sér. C*, 1, 87–94.

DANSON, D., BREBBIA, C.A. & ADEY, R. (1982). The BEASY system, *Adv. Eng. Software*, 4(2), 68.

EZBEA (1987). *EZBEA User Documentation, Version 2.2*, Caterpillar, Inc., Peoria, Illinois.

FLANIGAN, D.L. & BORDERS, S.G. (1983). Application of acoustic modeling methods for vehicle boom analysis. *Ford Motor Company, Automotive Eng.*, 207–17.

MANOLIS, G.D. & BANERJEE, P.K. (1986). Conforming versus non-conforming boundary elements in three-dimensional elastostatics, *Int. J. Numer. Methods Eng.*, 23, 1885–904.

NARDINI, D. & BREBBIA, C. (1982). Boundary integral formulation of mass matrices for dynamic analysis, *Proc. 5th Int. Conf. on BEM*, University of Southampton, UK.

NASTRAN (1965). *The MacNeal–Schwendler Corporation, NASTRAN Theoretical Manual*, Los Angeles, California.

NEFSKE, D.J., WOLF, J.A. & HOWELL, L.J. (1982). Structural acoustic finite element analysis of the automobile passenger compartment: a review of current practice, *J. Sound Vibr.*, 80(2), 247–66.

OSAWA, T. & IWAMA, A. (1986). A study of the vehicle acoustic control for booming noise utilizing the vibration characteristics of trunk lid, *SAE Passenger Car Meeting and Exposition*, Dearborn, Michigan, 22–25 September.

RIZZO, F.J. & SHIPPY, D.J. (1977). An advanced boundary integral equation method for three-dimensional thermoelasticity, *Int. J. Numer. Methods Eng.*, **11**, 1753–68.

SHUKU, T. & ISHIHARA, K. (1973). The analysis of the acoustic field in irregularly shaped room by the finite element method, *J. Sound Vibr.* **29**, 67–76.

Chapter 8

ADVANCED STRESS ANALYSIS BY
A COMMERCIAL BEM CODE

H.J. Butenschön, W. Möhrmann and W. Bauer

Daimler-Benz AG, Stuttgart, FRG

SUMMARY

On the basis of the Boundary Element program family known as BETSY, the largely interactive DBETSY program system has been developed; this system meets the requirements of industrial stress analysis with respect to operating comfort and range of applications.

Some of the new developments in DBETSY are presented here. Three examples from the current range of applications, in particular the usage of the three-dimensional module, show where—due to better economy—the Boundary Element Method has been able to gain acceptance in an analysis environment dominated by the Finite Element Method.

8.1 INTRODUCTION

The acceleration of industrial development processes while maintaining a high level of reliability and economy causes a steady increase in the importance of computational analysis, i.e. solving technical problems by means of mathematical models.

Rapid development of the performance of computers as well as the availability of increasingly efficient calculation methods are always necessary prerequisites. Particularly the highly developed Finite Element (FE) codes have influenced the whole sphere of technical analysis in industry in the course of the past two decades.

These FE programs had been in use for some time when with the work of Rizzo (1967) and Cruse (1969) the direct formulation of the Boundary Element Method (BEM) was presented for applications in the area of linear elastostatics.

After first implementations of this method for plane and three-dimensional applications (mainly by Lachat, 1975) had proved the advantages of this method over existing Finite Element programs—at least for a specific range of applications—and after further investigations had extended the range of applications to thermoelasticity (Rizzo and Shippy, 1977) and axisymmetrical geometry (Mayr, 1975, 1976; Cruse et al., 1977; Mayr et al., 1980), BEM also came to be of interest for industrial calculation.

The decisive advantage of modelling only the surface of a component to create a geometrical model, and not the whole volume as with the Finite Element Method, was so tempting that industrial analysis departments also began to adopt these programs.

In the Federal Republic of Germany this applies particularly to the field of mechanical engineering. These first attempts were very promising, and development of the BEM program family BETSY (Boundary Element code for Thermoelastic SYstems) was started as early as 1976, initiated by the automotive industry.

In the years 1977 to 1983, these programs were developed at Munich University under the direction of Professor Kuhn; today they are documented in Drexler (1982), Neureiter (1982), Neureiter et al. (1982) and Kuhn and Möhrmann (1983).

This program family will be briefly presented here.

Since the users were accustomed to the comfort of handling Finite Element systems, a series of requirements had to be fulfilled for practical applications of Boundary Element codes, so some modifications had to be made to BETSY, and DBETSY arose (Möhrmann, 1984, 1987). The structure and use of DBETSY are described here. Three typical and therefore particularly important examples show applications where—starting from the theoretical design stage—it is possible to obtain reliable results in a comparatively short period of time. Moreover, new calculations are possible with the BEM that failed in the past with the FEM because of the high expenditure.

This presentation is followed by a description of the advantages and disadvantages of BEM applications in comparison to FEM applications, and by prospects for further fields of application.

8.2 THE BEM CODE BETSY

BETSY was developed in the course of joint industrial research commissioned by the Forschungsvereinigung Verbrennungskraftmaschinen eV (FVV, Frankfurt) at Munich University within Research Projects Nos 209 and 245.

Figure 1 shows the structure of BETSY. First, different modules were

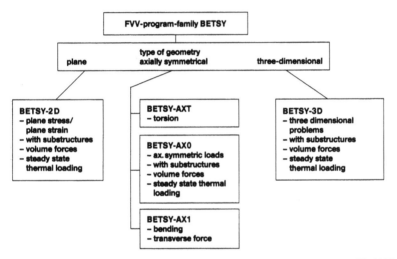

FIG. 1   Boundary Element code for thermoelastic systems (BETSY), 1977–1983.

developed depending on the type of geometry (plane, axisymmetrical or three-dimensional). In the case of axisymmetrical geometry and depending on the boundary value problem, three modules can be applied:

— AXT for torsion
— AX0 or AX1 in accordance with the 0th or 1st order of the Fourier expansion of the boundary values.

This constitutes a milestone, especially in the case of axisymmetrical geometry where it was possible to reduce three-dimensional problems to one-dimensional problems, particularly as initial experience already existed in the form of the results achieved by Mayr (1975).

The integration of the technique of substructuring and the possibility of including thermal effects or some volume forces was realised in the three modules 2D, AX0 and 3D.

The approximation possibilities that are integrated for discretising geometry, boundary values and integrals are specified in the documentation (Drexler, 1982; Neureiter, 1982; Kuhn and Möhrmann, 1983). The creation of the software was accomplished by a working group consisting of members from industry and research, and for this reason it was influenced by the requirements of industrial calculations at a very early stage. However, the primary development aim was to provide properly working codes, whereas the requirements made on operating comfort and possibilities of pre-processing and post-processing were only of secondary importance—particularly as they depend also on the data processing (DP) environment of a wide variety of future users.

The new modules were tested by some future users while still at the development stage. Radaj *et al.* (1984), for example, reported on economy and convergence in connection with the plane module, which—as early as 1979–showed obvious advantages of this BEM over an FE code, at least in the case of notch problems on surfaces of highly compact components.

A summary of the experience gained up to now and a wider theoretical context is given in Bausinger *et al.* (1987).

Since 1984 BETSY has been available on the market and is distributed by T-Program (Reutlingen) (Bausinger, 1984).

8.3 THE DBETSY PROGRAM SYSTEM

Depending on the user and the type of application, various requirements have been made on BETSY with regard to possible functions and handling. For this reason the integration of BETSY into the existing DP environment and the adaptation to the specific requirements should be carried out by the user himself on the basis of BETSY.

The program system DBETSY (<u>D</u>aimler-Benz <u>B</u>oundary <u>E</u>lement code for <u>T</u>hermoelatic <u>SY</u>stems) was thus developed at Daimler-Benz. This was done in two development stages:

— DBETSY 1.0 for all components which need only boundary curves to describe their geometry (in 1983)
— DBETSY 3D for three-dimensional applications, in 1985.

8.3.1 The Program System DBETSY 1.0
Figure 2 shows the structure of the program system (Möhrmann, 1984). It is

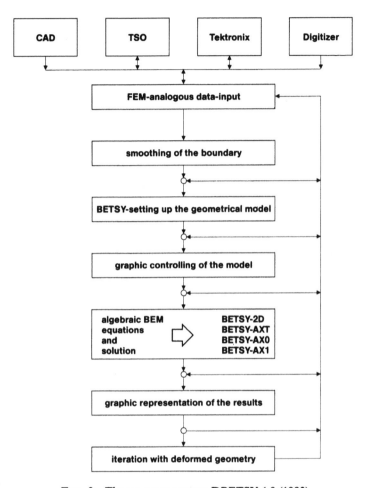

FIG. 2   The program system DBETSY 1.0 (1983).

possible to operate with DBETSY mostly in an interactive mode from terminals of various types; moreover, it is provided with interfaces to CAD programs and to a digitiser table so as to include this geometrical information as well.

The input data are specified analogous to an existing and frequently used Finite Element program with named data lines, and the geometry generation and check are concentrated in a first step which also provides graphical checks. Experience has shown that even small edges on the

boundary curves may have considerable influence on the stress distribution—at least in certain areas—and for this reason a smoothing pre-
processor which automatically smooths and rounds off undesired edges has
been developed and is offered as a part of the program system.

Once the geometrical model is created and tested and all boundary
conditions are specified, the basic BETSY programs are called up for setting
up and solving the relevant integral equations (either foreground or
background processing).

The stored and saved results can be graphically represented after
completion of this calculation. In the case of axisymmetrical geometry it is
possible—by connection to an FE data bus—to automatically attain
a three-dimensional representation of the results which is comparable to
the multiple FE post-processing possibilities.

In some cases, for example when analysing rubber components, the
simulation of geometrically non-linear behaviour with the help of BETSY
has also proved useful (Holzemer, 1982; Holzemer and Möhrmann, 1985);
for this purpose the deformation process is divided into several small steps,
and iterative analysis is carried out on the basis of the deformed geometry of
each preceding step; with this iteration non-linear rigidity curves can be
calculated with adequate accuracy. This process can be carried out with
DBETSY 1.0 in an automated mode.

In order to keep the expenditure of time and money within reasonable
limits, the possible number of grids per substructure has been limited to 200,
which is sufficient for the majority of applications.

8.3.2 The Program System DBETSY-3D

Figure 3 shows the structure of the program system for three-dimensional
applications. The aim is to make the system as similar as possible to the
existing Finite Element programs, in order to reduce the reluctance of new
users who are familiar with FE programs to use this new system. In addition
to an FE-oriented data input, this is ensured particularly by the connection
to the above-mentioned data bus. This makes it possible to exchange easily
both geometry and results between DBETSY and the available Finite
Element pre-processors and post-processors, where suitable. For this
purpose the geometry and results are stored in a binary mode by means of
data elements defined independently of special FE and BE codes (binary
input file (BIF) for geometrical elements, binary output file (BOF) for the
results). The pre-processing and post-processing programs have access to
this BIF/BOF data bus (Haase, 1984), just like DBETSY itself.

Thus, the geometry generation part which is carried out first provides all

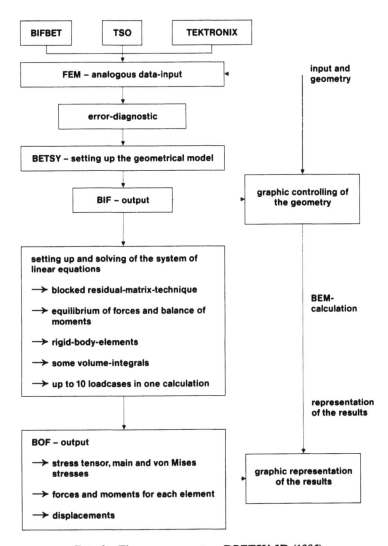

FIG. 3 The program system DBETSY-3D (1985).

the possibilities of DP-supported manipulation and representation that are also provided by FE applications.

Equally, the subsequent block for the representation of results also offers all the possibilities known from FE applications; apart from the familiar three-dimensional hidden-line representation, coloured or animated representations are possible, depending on the type of terminal available.

Especially for three-dimensional applications, the widest possible variety of representations of structure and results is of utmost importance because only in this way is it possible to convey the calculated results to the design or test engineer in such a way that misunderstandings and misinterpretations are avoided as far as possible.

In the geometry generation part, certain volume integrals which are of interest are calculated (foreground processing). Figure 4 shows the various quantities and the calculation by means of the Gaussian integral theorem.

The actual calculation of the generation and solution of the BEM integral equation system is carried out in the batch mode, which is indispensable due to the rather long CPU times for three-dimensional calculations. It is possible to solve up to 10 load conditions within one calculation process without significantly increasing the CPU time, since the vector on the right-hand side may have up to 10 columns. However, of course, only those load conditions can be solved together which may use the same equation matrix.

Additional possible boundary conditions which are integrated and

GAUSS: $\int\limits_{\Omega} \partial u/\partial x \, d\Omega = \int\limits_{T} u \cos{(n, x)} \, dT$

a) volume:
$$u = x \longrightarrow \text{mass}$$

b) coordinates of the centre of gravity (a, b, c):
$$u = 0.5 \, x^2$$

c) moment of inertia (related to the centre of gravity):
$$u = x \, [(y-b)^2 + (z-c)^2]$$

d) product of inertia (related to the centre of gravity):
$$u = x \, (y-b) \, (z-c)$$

FIG. 4 DBETSY-3D: calculated volume integrals.

available are the rigid body elements which are familiar from FE applications. For every substructure, the global balance equations (forces and moments) are checked and printed—this is an important indicator for possible problems with regard to the accuracy of the results.

8.3.3 The Blocked Residual Matrix Technique with DBETSY-3D

When carrying out the first three-dimensional calculations with DBETSY, one essential realisation was that it was practically impossible to describe real components adequately because for reasons of storage space the number of nodes per substructure was limited to 284.

Therefore, it was necessary to increase this number; this, however, would have led to enormous storage space problems due to the resulting large total matrix with the BETSY organisation of equation processing outlined in the upper part of Fig. 5.

BETSY–3D:

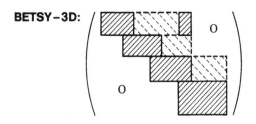

all degrees of freedom included,
max. 284 grids/substructure, blocks in core
or only one block in core

DBETSY–3D:

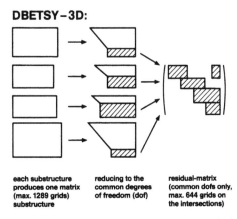

| each substructure produces one matrix (max. 1289 grids) substructure | reducing to the common degrees of freedom (dof) | residual-matrix (common dofs only, max. 644 grids on the intersections) |

FIG. 5   DBETSY-3D: residual matrix technique.

In accordance with a suggestion made by Bozek *et al.* (1983), the organisation of the overall equation system was broken down. The lower part of Fig. 5 illustrates the fact that now every substructure is first of all attached to its own part-matrix, which by means of forward substitution is then reduced to the degrees of freedom it has in common with other substructures. This eliminates those degrees of freedom which occur in this specific substructure only. The individual residual matrices (hatched) are combined to form the residual matrix. This reduced equation system is solved and subsequently the residual partial equation systems are also solved, by means of backward substitution.

It has proved practical, also due to data-processing properties, to use up to 1289 nodes per substructure with this technique, but even up to this stage it is not possible to store a complete part-matrix in core. Therefore, it was additionally necessary to split all matrix operations into block operations with line blocks. Figure 6 shows the forward substitution as an example of block technique. With this technique, two line blocks are always stored in core. The available size of the working-storage section determines the size (which can be specified) of the individual line blocks. To run the blocks, the second line block is changed until all lines have been processed. Then the next first line block is read in and the procedure is repeated. With this method, the column pivoting can only be carried out on the basis of the lines

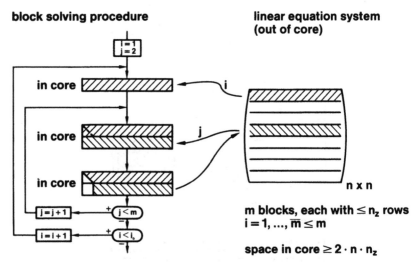

FIG. 6   DBETSY-3D: block-solving technique for forward substitution of linear BEM equations.

present in core. Therefore, any unnecessary infringement of the diagonal dominance must be avoided and will be corrected automatically in DBETSY-3D. If, however, pivoting problems do occur, the solution process starts anew with the former first line block now being shifted to the end of the equation system.

8.4 EXAMPLES OF APPLICATION

Since the introduction of DBETSY 1.0 (1983) and DBETSY-3D (1985), the Boundary Element Method as a supplement for the Finite Element Method reached a share of about 10% of all stress calculations at Daimler-Benz in 1986, with the FE applications including non-linear and dynamic applications for which DBETSY cannot be used in its present stage of development. With the increasing number of applications, expectations have been confirmed that the advantage of this method, constituted by the fact that less effort is needed for geometrical description as compared to the FEM, results in its being more economical in many cases than the FEM (refer also to Möhrmann (1987) and Möhrmann and Bauer (1987)).

The expected high accuracy of the results is particularly noticeable with compact components, the check of the balance equations by substructures having proved to be an essential criterion for indications of disturbances.

8.4.1 Part of a Camshaft

Figure 7 shows the Boundary Element models of the axisymmetrical geometry. On the left side, the half axial section with the given boundary curves and two substructures is plotted; by means of arc and straight elements it can be rapidly entered interactively at a terminal. The nodes for which boundary values are given or calculated are indicated by small subdividing lines on the boundary curves. Only part of these nodes must be given explicitly (two nodes for each line element, three nodes each per arc element), the rest of the nodes being generated automatically on the basis of the prescribed number of divisions.

In the case represented here 58 nodes and 48 elements had to be explicitly specified.

The three-dimensional model on the right-hand side has been generated automatically on the basis of the axial section with a division of 45° being given, using square isoparametric surface elements.

Figure 8 shows the deformed structure as a dotted line (enlarged by a factor of 1000) over the basic three-dimensional model as a first

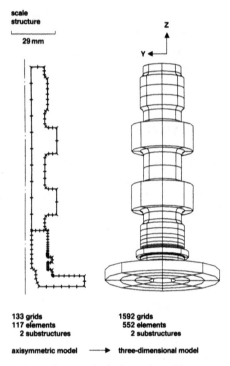

scale
structure

⊢——————⊣
29 mm

Z

Y

133 grids
117 elements
  2 substructures

1592 grids
552 elements
  2 substructures

axisymmetric model   ⟶   three-dimensional model

FIG. 7   Part of a camshaft, Boundary Element models.

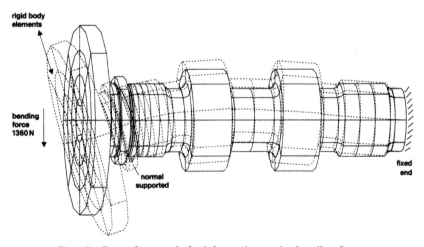

rigid body
elements

bending
force
1360 N

normal
supported

fixed
end

FIG. 8   Part of a camshaft, deformation under bending force.

representation of the results. The camshaft was subjected to bending on the left side at the sprocket via rigid body elements by applying a chain force and supported as specified. Despite axisymmetrical geometry, genuine three-dimensional analysis is necessary due to the type of one-sided support.

The matrix of the left-hand substructure contains about 8·88 million elements and the matrix corresponding to the right-hand substructure contains approximately 4·76 million elements. CPU time on a Cray-X/MP was about 31 min.

In a second model stage, the camshaft was provided with a sealing ring groove. Analogous to Fig. 7, Fig. 9 shows on the left-hand side the half-axial section and on the right-hand side the three-dimensional BEM model generated on the basis of this section.

As a first approximation, the application of the chain force can also be

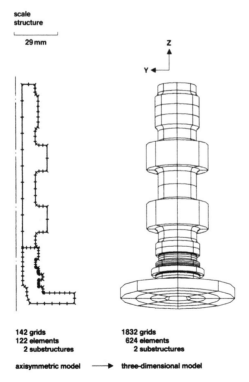

FIG. 9   Part of a camshaft with groove, Boundary Element models.

calculated with DBETSY 1.0 (AX1 module) if instead of one-sided support the whole circumference next to the groove is normally supported.

Figure 10 shows the result of such a calculation represented in the axial section of the deformed structure—plotted as a dotted line according to the given scale—over the axial section of the initial structure (represented by a continuous line), the whole model being described with one substructure.

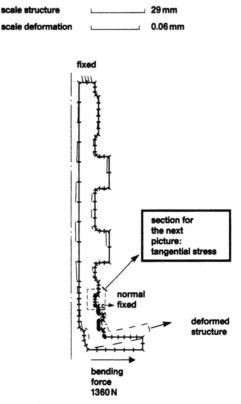

FIG. 10   Part of a camshaft with groove, deformation under bending force.

The calculated distribution of the tangential stress in the plane of projection has only been shown for the groove section (Fig. 11) for the sake of convenience. The stress values are plotted normally node by node inwards to the boundary curve, as they result for the axial section. They have a cosinusoidal distribution over the circumference (perpendicular to

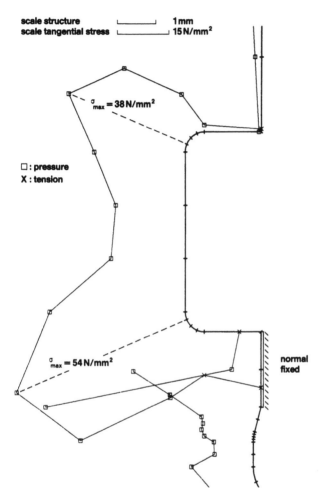

FIG. 11   Section of the camshaft model, tangential stress under bending force in the groove (AX1 calculation).

the plane of projection). In Fig. 11, the stress concentration in the groove notches is represented somewhat angularly because there are relatively few nodes in the notches, but the maximum values are already attained to about 90%, as was shown by a further simple convergence test with a finer mesh. Especially in the lower area of the support, a high pressure can be seen which suddenly jumps to a lower value at the corner. This is because only three nodes and linear elements have been used for the region of the support, and it is also due to the double-node concept used in BETSY for corners where the continuity of the stress tensor turning round the corner is used (analogous to Chaudonneret, 1977).

The CPU time for this AX1 calculation is so low that the calculation can be performed in the foreground; the matrix of the associated linear equation system contains about 0·174 million elements.

The surface stress (von Mises) in the case of genuine one-sided support—and thus as the result of a three-dimensional calculation for the analogue problem—is indicated in Fig. 12, viewed in the direction of the force perpendicular to the plane of projection, by means of iso-lines.

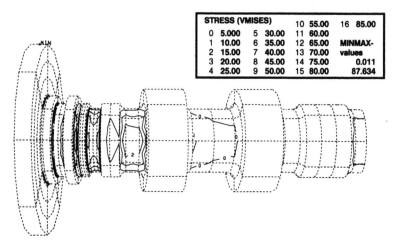

FIG. 12   Part of a camshaft with groove, stress distribution (von Mises).

Figure 13 shows a section of Fig. 12 with the groove area. It can be seen that due to the modified support, the difference between the stress peaks on the two notches, which are quite considerable in Fig. 11, has become smaller. The higher stress corresponding to Fig. 11 has decreased by about 20%, whereas the second stress peak has increased by about 8% (von Mises stresses).

Further investigations with regard to such a groove by means of DBETSY and its relief by means of relief grooves are contained in Bausinger et al. (1987).

In a third stage, oil ducts were included in the model, and as a result the geometric axisymmetry is lost. To illustrate this, Fig. 14 shows a section

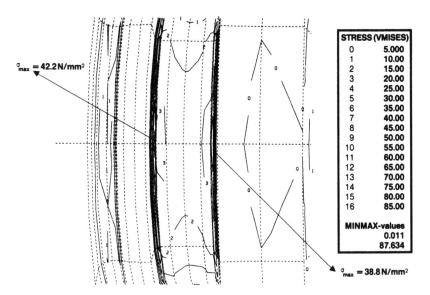

FIG. 13   Section of the camshaft model around the groove, stress distribution (von Mises).

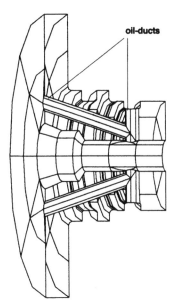

FIG. 14   Front part of the camshaft model, halved to show the oil ducts, Boundary Element model part.

through the first (left-hand) substructure carried out in such a way that the new oil ducts can be seen in full length.

Figure 15 shows the total model with ducts, which also consists of two substructures; the larger matrix of the first substructure contains about 13 million elements. Total CPU time on the Cray-X/MP increases to approximately 47 min.

The dotted line in Fig. 16 represents the structure deformed under bending force plotted against the basic structure; the deformations have

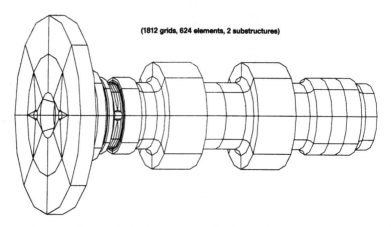

FIG. 15   Part of a camshaft with groove and oil ducts, Boundary Element model.

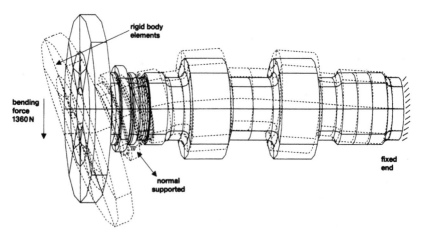

FIG. 16   Part of a camshaft with groove and oil ducts, deformation under bending force.

again been enlarged by a factor of 1000, and Fig. 17 shows the surface stress calculated according to von Mises. The view has again been chosen to show the force acting perpendicular to the plane of projection.

The detail of Fig. 17 shown in Fig. 18 clearly illustrates the effect of the oil

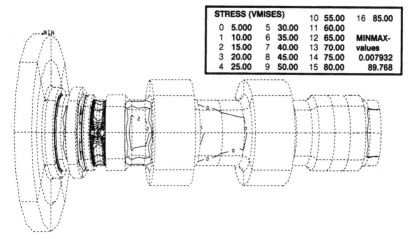

STRESS (VMISES)					
				10	55.00	16	85.00
0	5.000	5	30.00	11	60.00		
1	10.00	6	35.00	12	65.00	MINMAX-	
2	15.00	7	40.00	13	70.00	values	
3	20.00	8	45.00	14	75.00	0.007932	
4	25.00	9	50.00	15	80.00	89.768	

FIG. 17   Part of a camshaft with groove and oil ducts, stress distribution (von Mises) under bending force.

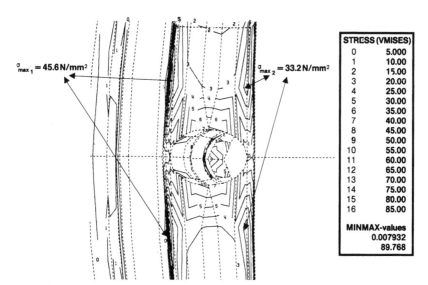

STRESS (VMISES)	
0	5.000
1	10.00
2	15.00
3	20.00
4	25.00
5	30.00
6	35.00
7	40.00
8	45.00
9	50.00
10	55.00
11	60.00
12	65.00
13	70.00
14	75.00
15	80.00
16	85.00
MINMAX-values	
	0.007932
	89.768

$\sigma_{max\,1} = 45.6\,N/mm^2$

$\sigma_{max\,2} = 33.2\,N/mm^2$

FIG. 18   Section of the camshaft model around the groove, stress distribution (von Mises) under bending force.

duct in the groove for this load condition. Due to the ducts, the stress peaks are shifted to the sides, and compared to Fig. 13 the stress peak has increased by about 8%.

In the various model stages, this example demonstrates the fact that on the basis of an initial axisymmetrical structure, which has been generated relatively quickly, complex three-dimensional structures (Figs 14 and 15), can be developed by means of relatively few modifications, and these structures can then be used for the calculations of different variants to assist the design and testing departments.

Moreover, developing the model via various model stages offers the advantage that the influence of the relevant modification on the overall component behaviour can be covered, which in turn substantially increases the understanding of the impact of individual design modifications.

8.4.2 Throw of a Crankshaft
Figure 19 shows a three-dimensional model of the throw of a crankshaft with balance weights and a bore (not directly visible here) in the crankpin; the relatively fine mesh in the longitudinal direction in the notches aims at achieving accurate stress results.

Analysing the stress concentration factors in the notches of the main journals and crankpins as well as the rigidity of the whole throw is one of the

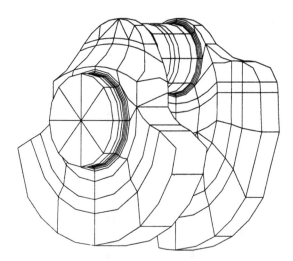

FIG. 19   Throw of a crankshaft with bore and balance weights, Boundary Element model.

main areas of application for DBETSY-3D (refer also to Bauer and Svoboda, 1985), since only with the introduction of the Boundary Element Method has it become possible to carry out such calculations within a short period of time and without much effort. Apart from the high accuracy of stress values in the notches (which has been demonstrated by appropriate tests), this is mainly due to the fact that a surface mesh generator suitable for such components has been developed. Thus, it is nowadays possible to obtain results with regard to stress concentration factors and rigidities only two or three hours after the theoretical design of such a throw.

The two substructures are represented separately in Fig. 20, which also includes the number of nodes and square isoparametric elements used. The corresponding two matrices of the resulting equation system contain about 5·9 million elements each; CPU time on a Cray-X/MP is approximately 27 min.

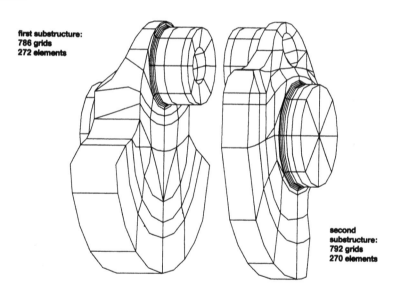

first substructure:
786 grids
272 elements

second substructure:
792 grids
270 elements

FIG. 20   Throw of a crankshaft with bore and balance weights—the Boundary Element model consisting of two substructures.

Figure 21 shows the deformed structure (dotted line) plotted over the basic structure (enlargement factor for the deformations is 150), when the two cut surfaces on the main journals are covered by rigid body elements. A torque of 1000 Nm is specified for the rotation round the y-axis on the

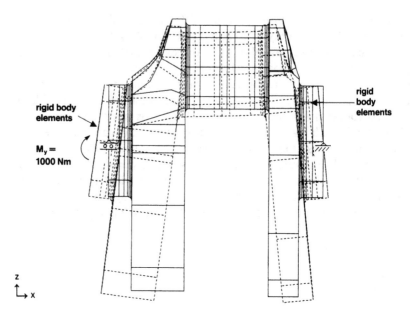

F<small>IG</small>. 21   Throw of a crankshaft, deformation (dotted line) under a bending moment.

left-hand side, and the rotational degree of freedom round the x-axis as well as the displacements of the central point of the cut surfaces are specified as zero on the right-hand side.

Figure 22 shows the stress distribution (von Mises) for this load condition by means of lines of constant stress on the surface of the deformed structure; for the sake of convenience, only the edges of the structure have been plotted. This lateral view clearly shows the stress-concentrating effect of the notches.

The stress concentration factors and rigidities thus calculated are input parameters in the subsequent calculations with other programs with regard to reliability or dynamic behaviour of the investigated crankshaft.

Besides the possibility of generating the BEM structure by means of the specific generator program, specifying only a few characteristic quantities, there is a second possibility, viz. using structures defined by means of CAD programs on the CAD side or on the calculation side for the generation of such BEM meshes.

This will be especially useful if new types of design are to be investigated for which the specific generator is not suitable.

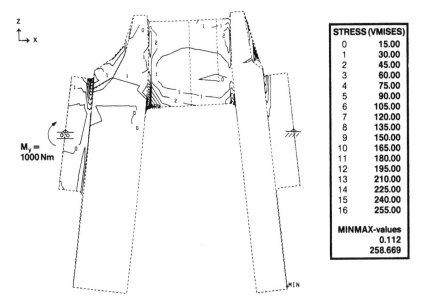

STRESS (VMISES)	
0	15.00
1	30.00
2	45.00
3	60.00
4	75.00
5	90.00
6	105.00
7	120.00
8	135.00
9	150.00
10	165.00
11	180.00
12	195.00
13	210.00
14	225.00
15	240.00
16	255.00
MINMAX-values	
	0.112
	258.669

FIG. 22   Throw of a crankshaft, stress distribution (von Mises) under a bending moment.

8.4.3 Bevel Gear Toothing

Figure 23 presents the three-dimensional BEM surface model of a bevel gear with only three of a total of 14 teeth explicitly modelled.

Each tooth corresponds to a substructure, and the cone body is additionally subdivided into three annular substructures arranged one after the other as shown in the exploded view in Fig. 24.

The equation systems corresponding to these substructures contain between 1·12 and 5·03 million elements; CPU time on the Cray-X/MP is approximately 22 min.

Figure 25 shows the deformed structure (dotted line) for the specified boundary conditions plotted over the initial structure; the deformations have been enlarged by a factor of 75.

The simulated stress on the medium tooth consists of constant normal stress at the tooth flank over the eight hatched surface elements, the gear being fixed at the hub.

Figure 26 is a detail showing the calculated stress distribution (von Mises) in the root of the tooth for this load condition; the notch effect of the root as well as the increase to the rear can be clearly seen.

DBETSY−3D-model:

2334 grids
881 elements
6 substructures

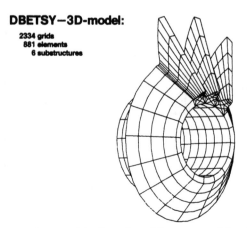

FIG. 23   Bevel gear (rear axle), Boundary Element model with three teeth explicitly modelled.

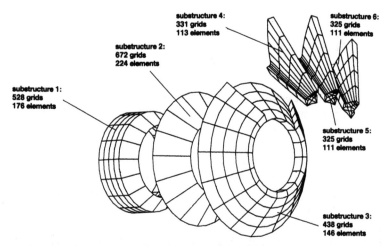

substructure 4:
331 grids
113 elements

substructure 6:
325 grids
111 elements

substructure 2:
672 grids
224 elements

substructure 1:
528 grids
176 elements

substructure 5:
325 grids
111 elements

substructure 3:
438 grids
146 elements

FIG. 24   Bevel gear (rear axle)—the Boundary Element model consisting of six substructures.

It is necessary to calculate on the basis of simulated loads because there is no information on the actual conditions of contact. Treatment of this contact problem of gearing teeth, including the model of the opposite tooth, and the iterative contact calculation is envisaged in the further development of DBETSY.

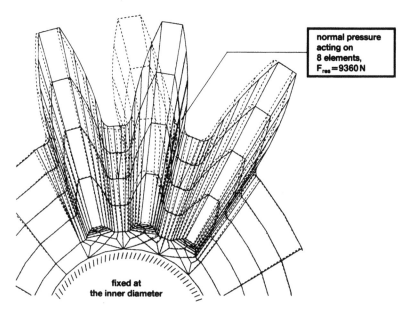

fixed at
the inner diameter

FIG. 25   Bevel gear (rear axle), deformation when forcing a tooth.

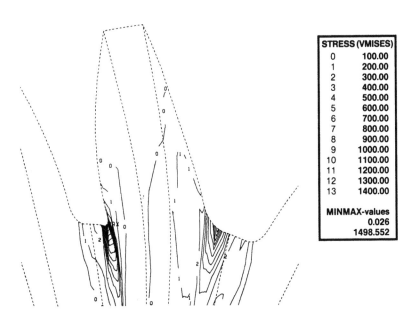

STRESS (VMISES)	
0	100.00
1	200.00
2	300.00
3	400.00
4	500.00
5	600.00
6	700.00
7	800.00
8	900.00
9	1000.00
10	1100.00
11	1200.00
12	1300.00
13	1400.00
MINMAX-values	
	0.026
	1498.552

FIG. 26   Section of the bevel gear, stress distribution (von Mises) in the root of the
teeth.

8.5 COMMENTS AND CONCLUSIONS

This paper presents the program system DBETSY with its two development stages DBETSY 1.0 and DBETSY 3D, which—on the basis of the BETSY program family—is used for industrial stress calculations within the framework of linear elastostatics.

Three examples with large numbers of degrees of freedom illustrate its application to complex three-dimensional components; the relatively high CPU times are kept within reasonable limits by means of a supercomputer of the Cray-X/MP type and a suitable solution strategy.

Further investigations will deal with further reducing the CPU times by increasing the degree of vectorisation, using parallel processors and iterative or mixed solution techniques.

The examples presented in this paper deal with notch problems on the surfaces of components which are relatively compact or which may at least be subdivided into relatively compact substructures. Experience has shown that for such problems the BEM provides advantageous possibilities of application compared to other finite methods, especially the FEM.

Figure 27 compares the essential characteristics of the BEM and the

BEM	FEM
• exact solution inside for integral kernel	• approximative solution inside
• only the boundary has to be described, low number of dofs	• the whole volume has to be described, high number of dofs
• high accuracy at compact parts, notches and stress concentrations, less accurate at non-compact parts, simple tests of convergence	• good accuracy also at branched parts, plates, bars • difficult test of convergence
• less expenditure of data pre- and postprocessing, simple mesh-generators for special parts	• high expenditure of data pre- and postprocessing, also with mesh-generators
• linear thermoelasticity, steady state heat conduction, acoustics	• wide range of application
• nearly 10% of all stress calculations at Daimler-Benz today	• nearly 90% of all stress calculations at Daimler-Benz today

FIG. 27   The BEM completes the FEM.

FEM so that the user can decide which of the methods is more suitable for his purposes. Because of the desired high degree of similarity between DBETSY and the existing Finite Element programs in terms of data handling and connection to the existing BIF/BOF data bus, there is no need for the user to feel reluctant about using DBETSY.

The lack of accuracy of the Boundary Element Method with non-compact components, such as beams, plates or shell-type parts, as described, for example, by Bozek et al. (1984), can also be observed with DBETSY. On the basis of experience gained so far, we recommend the carrying out of convergence tests, for example by refining the meshes, in order to check the quality of the results, in the case of dimension ratios from 10:1 upwards (largest:smallest component dimension). As described, the global balance equations mostly provide indications of problems via the substructures. An automated correction in cases where these global balance equations are poorly fulfilled is conceivable, for example by including the global balance equations in the Boundary Element equation system. Such investigations have already started.

Independent of such problems, it has been demonstrated that in the field of automotive engineering a variety of components can be calculated in an advantageous way by means of DBETSY. Extending such applications as suggested by, for example, Holze (1980) or Bozek et al. (1981), the examples put forward by Möhrmann (1984, 1987), Bauer and Svoboda (1985) and Möhrmann and Bauer (1987) illustrate this broad spectrum of applications.

So far, BETSY and DBETSY have been limited to stress and deformation calculations within the framework of linear elastostatics. The solution of the potential problem which can be carried out by means of analogue techniques has in the meantime been realised with the program code BETTI (refer also to Kuhn et al., 1987) so that now steady-state thermal conduction can also be dealt with by means of the BEM. BETTI was developed between 1984 and 1987 at the University of Erlangen–Nürnberg under the direction of Professor Kuhn at the request of the Forschungs-kuratorium Maschinenbau (FKM, Frankfurt) within Research Project No. 89.

Figure 28 shows the structure of BETTI. In accordance with the BETSY family, different modules were developed depending on the type of geometry (plane, axisymmetrical and three-dimensional). Each of these modules solves the problem of steady-state heat conduction for the specific type of geometry based on the Boundary Element technique, and its structure is very analogous to the corresponding BETSY module.

The resulting temperatures and heat-fluxes can be used as direct input data to BETSY together with the same geometrical model. An axisymmetric

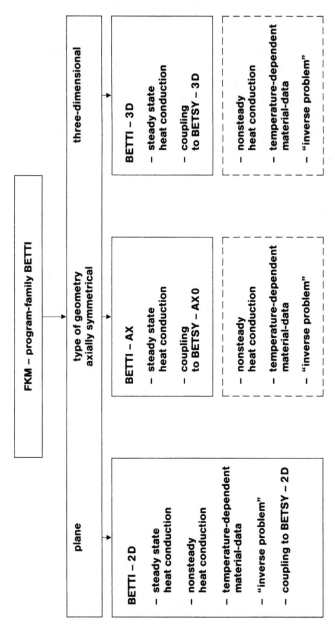

FIG. 28   Boundary Element code for heat conduction (BETTI), 1984–1987.

example for a BETTI–BETSY coupled calculation is given by Möhrmann (1987).

In the case of plane geometry, the project has comprised the investigation of the 'inverse problem' (replacing missing boundary values by interior data), temperature-dependent material data and the non-steady heat conduction, especially with regard to obtaining efficient codes compared with existing Finite Element programs. As a result, Kuhn *et al.* (1987) give examples for strongly ill-conditioned inverse problems. Moreover, the investigations of the different solution techniques for plane non-steady heat conduction have shown—with known BEM techniques—that it seems to be impossible to generate an efficient and generally applicable solution procedure based on boundary information only, so obtaining the BEM advantage over the Finite Element Method. Because we want to keep this main advantage of the BEM, the projected parts in the dotted boxes of Fig. 28 have not yet been realised. The BETTI program modules will be integrated into DBETSY in the near future.

REFERENCES

BAUER, W. & SVOBODA, M. (1985). Industrial application of the three-dimensional Boundary Element Method (BEM) exemplified through the BE program system DBETSY-3D, in: *Boundary Elements VII*, Eds C.A. Brebbia and G. Maier, Springer–Verlag, Berlin and New York, 14.3–14.20.

BAUSINGER, R. (1984). Boundary Element Methode: ein leistungsfähiges Berechnungsverfahren als Alternative zu FEM, *CAE-Journal*, No. 5, 28–31.

BAUSINGER, R., KUHN, G., MÖHRMANN, W., BAUER, W. & SEEGER, G. (1987). *Boundary-Element-Methode: Theorie und industrielle Anwendungen*, Expert–Verlag, Sindelfingen, FRG.

BOZEK, D.G., KATNIK, R.B., PARTYKA, J.K., KOWALSKI, M.F. & KLINE, K.A. (1981). Application of the boundary integral equation method (BIEM) to complex automotive structural components, *SAE paper 811321*, SAE, Warrendale, PA.

BOZEK, D.G., CIARELLI, D.M., HODOUS, M.F., CIARELLI, K.J., KLINE, K.A. & KATNIK, R.B. (1983). Vector processing applied to boundary element algorithms on the CDC CYBER-205, *EDF Bull. Dir. Etudes et Recherches, Sér. C*, 1, 87–94.

BOZEK, D.G., KLINE, K.A., KATNIK, R.B. & WOLF, W.L. (1984). Modeling considerations in use of Boundary Element Methods with substructures, *SAE paper 840735*, SAE, Warrendale, PA.

CHAUDONNERET, M. (1977). Calcul des concentrations de contraite en elasticité et viscoplasticité par la méthode des équations intégrales, *Int. Symp. Innovative Numerical Analysis in Applied Engineering Science*, Versailles, France.

260   H.J. BUTENSCHÖN, W. MÖHRMANN AND W. BAUER

CRUSE, T.A. (1969). Numerical solutions in three dimensional elastostatics, *Int. J. Solids Struct.*, **5**, 1259–74.

CRUSE, T.A., SNOW, D.W. & WILSON, R.B. (1977). Numerical solutions in axisymmetric elasticity, *Comput. Struct.*, **7**, 445–51.

DREXLER, W. (1982). Ein Beitrag zur Lösung rotationssymmetrischer thermoelastischer Kerbprobleme mittels der Randintegralgleichungsmethode, Dissertation, TU München.

HAASE, E. (1984). *Ein Weg zur integrierten Informationsverarbeitung in der Berechnung*, VDI–Tagung Fahrzeugtechnik, Fellbach, FRG.

HOLZE, G.H. (1980). Boundary integral equation method simplifies elastic stress analysis, *SAE paper 800431*, SAE, Warrendale, PA.

HOLZEMER, K. (1982). Berechnungsverfahren für Gummi und deren Einsatz bei der Entwicklung von Gummibauteilen, *VDI-Ber.*, No. 444, 109–20.

HOLZEMER, K. & MÖHRMANN, W. (1985). Rechnerische Auslegung von Gummibauteilen mit der Boundary-Element-Methode (BEM), In *8. Reutlinger Arbeitstagung: Finite Element Methoden in der Praxis*, Fa. T-Programm, Reutlingen, FRG, pp. 202–3.

KUHN, G. & MÖHRMANN, W. (1983). Boundary element method in elastostatics: theory and applications, *Appl. Math. Model.*, **7**, 97–105.

KUHN, G., LÖBEL, G. & SICHERT, W. (1987). *Erstellung des Programms BETTI zur Berechnung der stationären und instationären Wärmeleitung mittels BEM*, Final report, Forschungsbericht des Forschungskuratoriums Maschinenbau (FKM), Frankfurt, FRG.

LACHAT, J.C. (1975). A further development of the boundary integral technique for elastostatics, Ph.D. Thesis, University of Southampton, UK.

MAYR, M. (1975). Ein Integralgleichungsverfahren zur Lösung rotationssymmetrischer Elastizitätsprobleme, Dissertation, TU München.

MAYR, M. (1976). On the numerical solution of axisymmetric elasticity problems using an integral equation approach, *Mech. Res. Commun.*, **3**, 393–8.

MAYR, M., DREXLER, W. & KUHN, G. (1980). A semianalytical boundary integral approach for axisymmetric elastic bodies with arbitrary boundary conditions, *Int. J. Solids Struct.*, **16**, 863–71.

MÖHRMANN, W. (1984). DBETSY—die Boundary Element Methode in der industriellen Berechnung, *VDI-Ber.*, No. 537, 627–50.

MÖHRMANN, W. (1987). Industrial application of the BEM using DBETSY, *IUTAM Symp. on Advanced BEM: Applications in Solid and Fluid Mechanics*, San Antonio, Texas.

MÖHRMANN, W. & BAUER, W. (1987). DBETSY—industrial application of the Boundary Element Method, in: *Boundary Elements IX*, Eds C.A. Brebbia, W.L. Wendland and G. Kuhn, Springer–Verlag, Berlin, vol. 1, 593–607.

NEUREITER, W. (1982). Boundary-Element-Programmrealisierung zur Lösung von zwei- und dreidimensionalen thermoelastischen Problemen mit Volumenkräften, Dissertation, TU München.

NEUREITER, W., DREXLER, W., MEWS, H. & MÖHRMANN, W. (1982). *Integralgleichungsmethode*, Final report I–III, Forschungsberichte Heft 310-1, -2, -3, Forschungsvereinigung Verbrennungskraftmaschinen eV, Frankfurt, FRG.

RADAJ, D., MÖHRMANN, W. & SCHILBERTH, G. (1984). Economy and convergence of notch stress analysis using boundary and finite element methods, *Int. J. Numer. Methods Eng.*, **20**, 565–72.

RIZZO, F.J. (1967). An integral equation approach to boundary value problems of classical elastostatics, *Qt. Appl. Math.*, **25**, 83–95.

RIZZO, F.J. & SHIPPY, D.J. (1977). An advanced boundary integral equation method for three-dimensional thermoelasticity, *Int. J. Numer. Methods Eng.*, **11**, 1753–68.

Chapter 9

THERMOELASTIC ANALYSIS FOR DESIGN
OF MACHINE COMPONENTS

GERALD L. GRAF and YOSEPH GEBRE-GIORGIS

Caterpillar Inc., Peoria, Illinois, USA

SUMMARY

Development and use of boundary element analysis at Caterpillar is discussed. Major issues in the development of Caterpillar's BE programme, including multi-subregion capability, are reviewed. Six examples of boundary element analysis of machine components are described, including compute times on various computers.

9.1 INTRODUCTION

9.1.1 The Role of Analysis in Product Development

Designers in many industries are finding that product development cycles can be shortened by introducing analysis early in the design process. This analysis reduces the number of prototypes that must be built and tested.

For the analysis to be effective, it must be timely—results must be available before designs are committed to production. The cost of a design change increases dramatically as the design moves from concept to layout to prototype to production. The most effective analysis provides information during the early stages of the design when changes can be made most easily.

If analysis is used in lieu of testing for product evaluation, the analysis results are generally expected to provide detail comparable to a full scale test. This generally requires large models that take longer to build and solve and produce more results which must be reviewed. This trend to larger, more detailed models, combined with the need for timely analysis, forces the analyst to seek analysis methods that require less time.

9.1.2 BEA vs FEA

Caterpillar designers, like others described in this book, have found that boundary element analysis (BEA) can be used instead of finite element analysis (FEA) to shorten the analysis time required for some components. However, timeliness is only one aspect to be considered in choosing an analysis method. For the industrial analyst comparing BEA and FEA, three considerations stand out:

(1) *Cost*. What are the relative manpower and computing costs for using the BEA and FEA programs being compared?
(2) *Accuracy*. For comparable models, what is the relative accuracy of the programs being compared?
(3) *Capability*. Does each program have the features required to solve the desired models?

Our experiences indicate that there are certain classes of models where BEA has a clear cost advantage over FEA. These are 'chunky' models with relatively small surface-to-volume ratios or complex geometry such as intersecting holes. The cost advantage is primarily due to time savings in preparing the model and reviewing the results. Though computing costs are more difficult to compare, BEA seems to have the advantage for small problems but to be at a disadvantage for large problems.

Accuracy comparisons between BEA and FEA are difficult to make for real problems. Most technical papers (e.g. Mukherjee and Morjaria, 1984) show that BEA is more accurate than FEA for comparable mesh refinement. As justification, authors such as Gerstle and Ingraffea (1985) have noted that only four approximations are generally made in the boundary element method:

(1) The geometry is approximated by elements.
(2) The boundary conditions are approximated by some specified variation over the boundary.
(3) The coefficient matrix is generated by numerical integration.
(4) The computer has a finite word size.

The finite element method has an additional approximation—the field variable(s) are approximated by some specified variation throughout the domain. On the surface, it seems clear that BEA has an accuracy advantage.

Our experience shows that this is true for 'good' models. However, our experience with 'poor' models (i.e. models with elements that are too large or are poorly shaped) is that, though both FEA and BEA produce inaccurate results, the FEA results may seem 'reasonable' while the BEA results are 'unreasonable'. The reader will have to decide for himself which has more value.

Today's commercial FEA programs have a clear capability advantage over today's BEA programs. We believe this is primarily due to the amount of development that has gone into each method. Our experience indicates that a secondary reason is the level of effort required to program each method, the FEM being inherently modular and easier to program. The FEM also has only one compute-intensive operation (matrix solution) while the BEM has two (matrix generation and matrix solution). This implies that less FEA code must be optimised to achieve acceptable performance.

9.1.3 Requirements for Analysis Software

The FEA programs available today are generally very good. Analysts accustomed to these FEA programs have high expectations for alternative analysis software, such as BEA programs. In particular, the analysts expect *reliability* and *performance*. By reliability, we mean that if the user enters valid input data, the program produces the correct answer. At a minimum, the following are required:

(1) The underlying mathematics must be correct.
(2) The algorithm used to implement the mathematics must perform correctly.
(3) The program implementing the algorithm must perform correctly.
(4) The computer system that runs the program must perform correctly.

The first and second are generally established through peer review in the technical literature. The software developer is responsible for selecting algorithms that work and for implementing the algorithms correctly ((3) above). The fourth requirement appears to be totally the responsibility of a computer operations group, though in reality the software developer can do much to produce a program that is compatible with the way in which the computer is operated. The program may need to be able to perform a large analysis in a piecemeal fashion, stopping occasionally for system backups.

9.2 EZBEA DEVELOPMENT

9.2.1 EZBEA History

The authors are two of the developers of the EZBEA program and, consequently, will outline its development here. Caterpillar's interest in boundary element problems began in 1977 when Dr Gordon Holze recognised the potential that BEA held for reducing the time required to build models of complex shapes (see Holze (1980)). He sought a commercially available BEA program but found the programs then available to be unsatisfactory. In 1979, Caterpillar agreed to purchase a licence to use the BEA programs developed by Dr Frank Rizzo and Dr David Shippy of the University of Kentucky. Parts of these programs were used in EZBEA.

The initial version of the EZBEA program became available to Caterpillar's engineering analysts in 1981. This version, which could model only a single region, used code developed by Rizzo and Shippy to generate the coefficient matrix and to recover results after the matrix solution. The solver was based on Gauss elimination and was similar to that described by Lachat (1975), Lachat and Watson (1976), and Watson (1976). Since the program was designed for solving 'real' problems which might require long solution times, special attention was given to finding modelling errors before the solution began.

A second version of EZBEA was released in 1984. This version could model up to 10 subregions. Model size limits were increased commensurate with the increased computer power available to the users. The capability to model non-axisymmetric boundary conditions on axisymmetric models was also included. A subsequent 'minor' release in 1985 contained several evolutionary changes to accommodate larger models. These changes included further increases in size limits and full restart capability during matrix generation and solution.

The current version of EZBEA, 2.2, was completed in 1986. This version's capabilities are summarised in Table 1. The development of this version is described further in the following section.

9.2.2 EZBEA Development Issues

EZBEA 2.2 was designed to solve large models in a 'production' environment. Four attributes were emphasised during the development—reliability, portability, efficiency, and compatibility.

9.2.2.1 Reliability

If the user enters valid input data, we expect the program to produce a valid answer. We do not expect our users to be boundary element 'experts';

TABLE 1
SUMMARY OF EZBEA 2.2 CAPABILITIES

Analysis types	• Steady-state heat conduction • Static thermoelasticity
Domain types	• Planar • Axisymmetric • General three-dimensional
Symmetry	• Domain symmetry and loading symmetry about X, Y, and/or $Z=0$ • Domain symmetry and loading antisymmetry about X, Y, and/or $Z=0$
Material	• Isotropic, linear; one material per subregion
Loading (Heat conduction)	• Specified surface temperatures • Specified surface fluxes • Surface film conductivity (convection) • Non-axisymmetric loading on axisymmetric models
Loading (Stress analysis)	• Temperatures and fluxes from heat conduction analysis • Specified uniform subregion temperature • Gravity loading • Centrifugal loading • Specified surface displacements in arbitrary directions • Specified surface tractions in arbitrary directions • Specified pressure loading • Specified elastic restraints in arbitrary directions • Non-axisymmetric loading on axisymmetric models
Pre-processing	• Determines subregion membership from element connectivity • Automatic two-dimensional grid generation and refinement
Post-processing	• Combine load cases • Plot two-dimensional analysis results • Interior point results • Neutral output file

rather, we expect a typical finite element analyst to be able to perform boundary element analyses with minimal training.

9.2.2.2 Portability
Our engineering analysts currently use APOLLO, DEC, CDC, IBM, and FPS computers. Many factors influence which computer is used to solve

a given model; we did not want software availability to be one of those factors. Consequently, EZBEA 2.2 generally utilises FORTRAN-77 (i.e. where there is a standard). The relatively small number of lines of non-standard code are automatically entered or deleted by a source code management system—a single source code is maintained. EZBEA also uses only three working files. These have fixed record lengths which the user can set at execution time. This further improves the portability of the program.

9.2.2.3 Efficiency

We have used software performance monitor tools to help us optimise the efficiency of the numerically intensive parts of the EZBEA program. We have also paid particular attention to the use of central memory and disc storage and can easily configure the program to best utilise the resources available. The program's main storage array can be easily redimensioned to best fit the physical memory that is available; disc storage is used to supplement the available memory.

9.2.2.4 Compatibility

A final design attribute for EZBEA 2.2 was compatibility. Our users are finite element analysts; our EZBEA program must fit into their finite element world. This implies that the pre-processing and post-processing programs used with FEA should also work with BEA. In our case, pre-processing is done using SUPERTAB® or a Caterpillar-developed program; post-processing is done using SUPERTAB®or FEMVIEW®. We designed EZBEA to facilitate linkage to these programs or other pre/post-processers and we have the requisite links in place at Caterpillar.

9.2.3 EZBEA Capability—Heat Conduction

For heat conduction problems (or others governed by Laplace's equation) the boundary integral equation (BIE) which relates the potential at any point P to the potential and the derivative of the potential at any point Q on the boundary of the domain, is given by the following equation:

$$C(P)\phi(P) + \int_S [A(P, Q)\phi(Q) - B(P,Q)\phi, n(Q)]\, \mathrm{d}S(Q) = 0 \qquad (1)$$

where

$\phi(P)$ is the potential function at point P,
$\phi(Q)$ is the potential function at point Q,

$\phi, n(Q)$ is the derivative of the potential function with respect to the normal direction at point Q, and $C(P)$, $A(P, Q)$, and $B(P, Q)$ are the kernel functions.

In EZBEA, parabolic elements are used to model the domain of interest; algebraic equations are generated by collocation of eqn (1) at corner and midside nodes. The elements are similar to those described by Rizzo and Shippy (1977), though a more accurate numerical integration scheme has been implemented.

In eqn (1), the potential function ϕ is continuous on the boundary S. The boundary variable $\phi, n(Q)$ is a function of both ϕ and the normal vector on the boundary; it may not be continuous between elements. The boundary integral equation provides one equation for each node on the boundary of the model. If the variable ϕ, n is specified, the unknown variable ϕ can be determined from the integral equation. However, if ϕ is specified on adjacent elements, and the normal vectors of these elements are not parallel, the variable ϕ, n is unknown on *both* elements. To resolve this, a 'parent element' is selected for each node. The $\phi, n(Q)$ unknowns of all elements connected to a node are expressed in terms of ϕ and $\phi, n(Q)$ of the parent element. The approach is similar to that described by Rudolphi (1983).

Both the normal flux $\phi, n(Q)$ on any element and the kernel function $B(P, Q)$, which multiplies $\phi, n(Q)$, can be expressed in terms of the normal and parallel directions on the 'parent element'. This transformation may be between elements in the same subregion or in different subregions. The transformation of $\phi, n(Q)$ to a parent element in the same subregion is a rotation of coordinates based on components of the normal and parallel vectors. The transformation between elements in different subregions requires equations which relate the derivatives of the potential across the interface between the subregions. This transformation is based on the following assumptions:

(1) The derivatives of the potential in the directions parallel to the element are continuous across the subregion interface.
(2) The flux normal to the element's surface is equal and opposite in the two subregions.

Performance of EZBEA for many heat conduction problems, including models with multiple subregions and discontinuous normals, is documented in the *EZBEA Verification Examples Manual* (Anon., 1986).

9.2.4 EZBEA Capability—Thermoelasticity

The boundary integral equation for thermoelastic analysis is a constraint equation which relates the displacements at point P to the displacements and tractions at every point Q on the boundary of the domain. Equation (2) shows a simplified form of the BIE for elasticity which neglects thermal and body forces.

$$C_{ij}(P)u_j(P) + \int_S [u_i(Q)T_{ij}(P, Q) - t_i(Q)U_{ij}(P, Q)]\, \mathrm{d}S(Q) = 0 \qquad (2)$$

where

$u_j(P)$ is the displacement in the j-direction at point P,
$u_i(Q)$ is the displacement in the i-direction at point Q,
$t_i(Q)$ is the traction in the i-direction at point Q, and
c_{ij}, U_{ij}, T_{ij} are the kernel functions.

As for heat conduction, EZBEA uses parabolic elements to model the domain of interest; algebraic equations are generated by collocation of the BIE at corner and midside nodes. The elements are similar to those described by Rizzo and Shippy (1977), though the numerical integration is improved.

In eqn (2), displacement is a nodal variable and is continuous. Traction is a variable associated with elements and may not be continuous; a node may have various tractions associated with the elements that share it. In EZBEA, each node has at most three unknowns in each of the subregions in which it appears. For single subregion models, either the displacement or the traction is known in each direction at each node; the program solves for the other. When displacements are specified on adjacent elements, the tractions are unknown on each element in the direction of the specified displacements. Under such conditions, the number of traction unknowns may exceed the number of equations provided by the BIE.

As in heat conduction, the concept of a 'parent element' for a node is used to resolve this problem of too many unknowns and not enough equations. Transformation matrices are developed to relate tractions on any element connected to a node to tractions and strains on the parent element of the node.

Performance of EZBEA for many elasticity problems, including models with multiple subregions and discontinuous normals, is documented in the *EZBEA Verification Examples Manual* (Anon., 1986).

9.2.5 Multiple Subregions—Implementation

EZBEA can be used to solve two-dimensional and three-dimensional multi-subregion problems in steady-state heat conduction and static thermoelasticity. The number and types of unknowns per node depend on the connection of subregions at the node. Multi-subregion models may contain the following types of nodes:

(i) nodes found on a surface;
(ii) nodes found on an interface between two subregions;
(iii) nodes found on both an interface and a surface; and
(iv) nodes found in the interior, shared by three or more subregions.

Figure 1 shows a two-dimensional model of a three-subregion problem; it shows the various types of nodes.

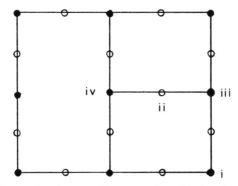

FIG. 1   Three-subregion two-dimensional model showing node types.

9.2.6 Multiple Subregions—Implementation (Heat Conduction)

Table 2 shows the types of unknowns in heat conduction for the various node types found in multi-subregion models. Note that n, p, and q are the normal, primary parallel, and secondary parallel directions on the 'parent' element. For surface nodes (node type i), either the temperature or the flux $(-k\phi, n)$ or a linear combination of the two is specified as boundary conditions. The unspecified variable is the unknown.

At interface nodes shared by two subregions (node type ii) both the temperature and its normal derivative are unknown. Since the node is shared by two subregions, each contributes an independent integral equation. Thus, the temperature is unknown in one subregion while the normal derivative of the temperature is unknown in the other subregion.

GERALD L. GRAF AND YOSEPH GEBRE-GIORGIS

TABLE 2
UNKNOWNS IN MULTI-SUBREGION HEAT CONDUC-
TION PROBLEMS

Node type	Types of unknown variables
i	ϕ or ϕ,n
ii	ϕ and ϕ,n
iii	ϕ or ϕ,n; ϕ,p and ϕ,q
iv	ϕ; ϕ,n; ϕ,p and ϕ,q

Interface surface nodes (node type iii) are both surface nodes (with specified boundary conditions) and interface nodes (shared by two or more subregions). Since they are interface nodes, they have unknowns which are used to define the heat flux across the subregions. The equations for the temperature unknown (ϕ) or normal derivative of temperature (ϕ,n) are provided by one subregion while the equation of ϕ,p is provided by another. The ϕ,n's of all adjacent elements are transformed into components of ϕ,n and ϕ,p of the 'parent element' of the node.

Interface interior nodes (node type iv) shared by three or more subregions are similar to interface surface nodes with one exception: they are on the interior and no boundary conditions are specified. Thus, both ϕ and ϕ,n are unknown. Also, the variables which relate the heat flux across the subregions are unknown. One subregion provides an integral equation for ϕ; a second subregion provides an integral equation for ϕ,n; a third subregion provides an integral equation for ϕ,p.

9.2.7 Multiple Subregions—Implementation (Thermoelasticity)

Table 3 shows the types of unknowns in thermoelasticity for the various node types found in multi-subregion models. As previously stated, surface

TABLE 3
UNKNOWNS IN MULTI-SUBREGION THERMOELASTICITY
PROBLEMS

Node type	Types of unknown variables
i	u_i or t_i $(i=1,2,3)$
ii	u_i and t_i
iii	u_i or t_i, σ_p, σ_{pq}, and σ_q
iv	u_i, σ_n, σ_{np}, σ_{nq}, σ_p, σ_{pq}, and σ_q

nodes (node type i) have either displacements or tractions or a linear combination of the two specified as boundary conditions. Therefore, the unspecified value becomes the unknown.

Interface nodes shared by two subregions (node type ii) have both displacement and traction unknowns. In this case the BIE provides one independent equation for each degree of freedom in each of the subregions.

Interface surface nodes (node type iii) have specified boundary conditions and additional unknowns which are used to define the states of stress at the nodes. The equations for the displacement or traction unknowns are provided by one subregion. Traction variables on elements not parallel to the 'parent element' are written in terms of the stresses σ_p, σ_{pq}, and σ_q on the 'parent element' of the node. The tractions and these stresses are sufficient to define completely the stress state at the node.

Interface interior nodes shared by three or more subregions (node type iv) are similar to interface surface nodes except that they have no boundary conditions. For three-dimensional analysis, these nodes have equations for the three displacements provided by one subregion. The tractions on any of the elements which share the node are transformed into a stress tensor in the normal, primary parallel, and secondary parallel (n, p, q) directions on the 'parent element'. The first three stress unknowns are written for a second subregion while the last three are written for a third subregion.

9.2.8 Multiple Subregions—Heat Conduction Example

The following example is used to demonstrate the steady-state heat conduction analysis capability of EZBEA for problems with non-homogeneous domains. The problem involves one-dimensional heat flow through a two-material cylinder. The dimensions of the problem are shown in Fig. 2. The inside surface is maintained at a constant temperature of 100°C, while the outside temperature $T3$ is 0°C. The coefficients of thermal conductivity for materials A and B are 3·0 and 2·0 W/mm^2 °C respectively.

The problem was solved using the three models (planar, axisymmetric, and three-dimensional) shown in Fig. 2. The objective was to determine the interface temperature $T2$ and the heat flux across the interface. The solution of the problem is summarised in Fig. 2 ($T2$ is the interface temperature in °C, Q is the heat flux in W/mm^2).

9.2.9 Multiple Subregions—Thermoelasticity Example

The example shown in Fig. 3 demonstrates EZBEA's multi-subregion thermoelastic analysis capability. This is a plane strain analysis of con-

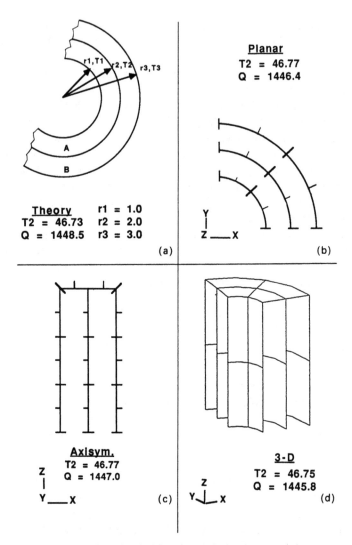

FIG. 2   Multi-subregion heat conduction example.

centric cylinders with interference fit. The inner cylinder's temperature increases 100°, while the outer cylinder's does not.

The planar model shown in Fig. 3(b) was used to solve this problem. The objective was to determine the interference stress between the cylinders. Figure 3(c) shows a plot of the stress in the radial direction along the x-axis

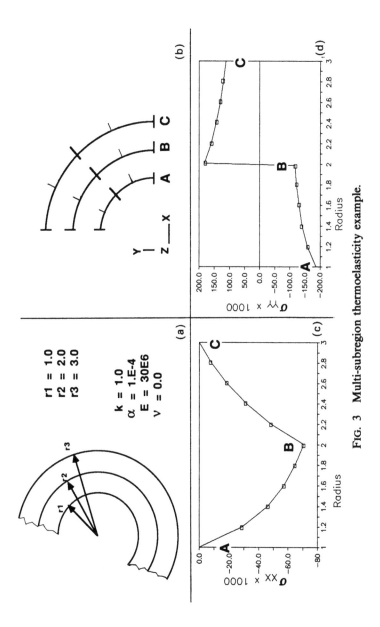

FIG. 3   Multi-subregion thermoelasticity example.

(the data are from nodes and interior points). The interference stress at the interface between the cylinders (point B) is predicted to be — 70 420; the solution given by Shigley (1972) is —70 310. Figure 3(d) shows a plot of the hoop stress along the x-axis.

9.2.10 Non-axisymmetric Loading of Axisymmetric Models

Problems with geometry that can be defined by a solid of revolution can be analysed using two-dimensional axisymmetric models. If the boundary conditions are axisymmetric, the field variables, such as displacements and stresses, are uniform along the circumferential direction. This type of problem can be solved by considering only radial and axial coordinates.

If the boundary conditions are not axisymmetric, the boundary variables and kernel functons in eqn (2) may be expanded in a Fourier series as follows:

$$u_i(P) = \frac{u_i^0(P)}{2} + \sum_{m=1}^{\infty} [{}^s u_i^m(P) \sin m\Theta + {}^c u_i^m(P) \cos m\Theta] \tag{3}$$

The expanded terms are substituted into the boundary integral equation, like terms are collected, and orthogonality of the sine and cosine functions is used to produce an independent boundary integral equation for each mode (m) such as the following (see Nigam (1980) for details):

$$C_{ij}(P)^s u_j^m(P) = \int_S [{}^s t_i^m(Q) U_{ij}^{mc}(P, Q) - {}^c t_i^m(Q) U_{ij}^{ms}(P, Q)$$

$$- {}^s u_i^m(Q) T_{ij}^{mc}(P, Q) - {}^c u_i^m(Q) T_{ij}^{ms}(P, Q)] r \, dS(Q) \tag{4}$$

where $m = 0, 1, 2, \ldots, \infty$.

In EZBEA, axisymmetric problems with non-axisymmetric loading are solved by treating each mode in the Fourier series as an independent load case. The coefficients of the series become the boundary variables. EZBEA can solve for problems which are characterised by the first five terms of the Fourier series (constant, sin Θ, cos Θ, sin 2Θ, cos 2Θ).These terms are sufficient to describe the loading of most practical non-axisymmetric problems.

9.2.11 Non-axisymmetric Loading of Axisymmetric Models—Example

The following example demonstrates the use of non-axisymmetric loading in axisymmetric problems. A stepped round bar shown in Fig. 4 is subjected to a moment load. In this example, a moment of 1000 is applied on the end

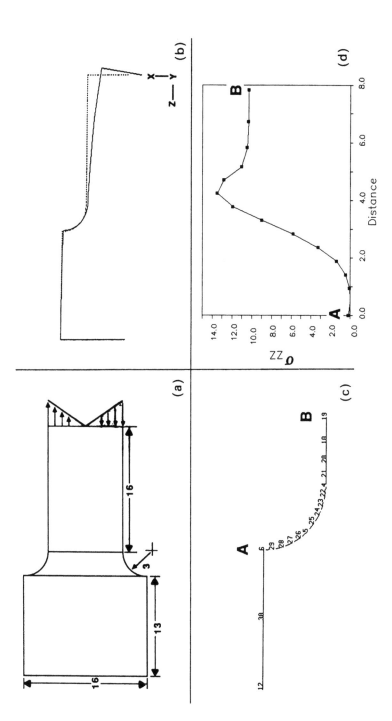

(a)

(b)

(c)

(d)

Fig. 4 Stepped round bar under bending load.

of the bar as shown in Fig. 4(a). Figure 4(b) shows the deformed shape for this problem. The axial stress along the radius shown in Fig. 4(c) is plotted in Fig. 4(d). The stress concentration factor given by Roark and Young (1975) for this problem is 1·3357; the value calculated by EZBEA is 1·3244.

9.2.12 Advanced Numerical Integration

Our experience has shown that the accuracy of the BEM can deteriorate rapidly as the model becomes 'thin' and the coefficient matrix becomes both inaccurate and poorly conditioned. In EZBEA 2.2, an advanced numerical integration scheme for the non-singular integrands has greatly reduced this potential inaccuracy.

In EZBEA 2.2, the integration of the non-singular integrands is done over subelements where the subelement size is based on how nearly singular the integrand is. A recursive algorithm similar to that described by Cox and Shugar (1985) is used to eliminate unnecessary computations while maintaining accuracy as the integrand becomes nearly singular. This algorithm has been shown to provide improved accuracy for 'thin' models and for interior points 'close' to the surface of the model. Figure 5 compares the performances of EZBEA 2.1B, which uses the integration scheme described by Rizzo and Shippy (1977), and EZBEA 2.2. The abscissa of the

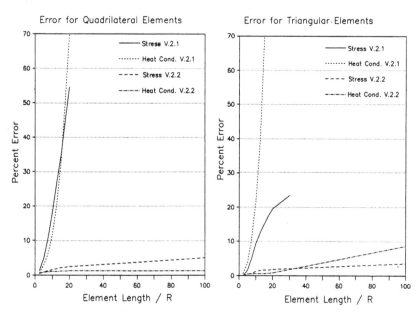

Fig. 5   Effect of element size/spacing ratio on accuracy.

plots is the ratio of the maximum element dimension to the spacing between discretised surfaces.

As a further demonstration of the accuracy of the new integration scheme, a cube model with discontinuous discretisation (see Fig. 6) was

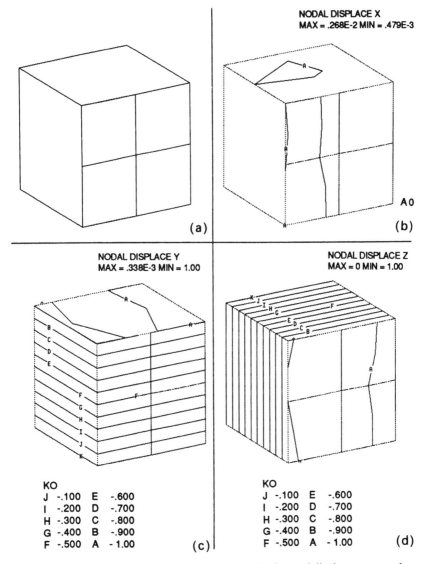

FIG. 6   Three-dimensional discontinuous discretisation and displacement results.

solved. Since EZBEA uses parabolic elements with corner and midside nodes, the model shown in Fig. 6 in fact used the advanced 'non-singular' integration scheme to integrate 'singular' integrands. The results (displacements and stresses) differ from the exact solution by less than 0·5% (x, y, and z displacement contours are shown in Fig. 6(b), (c) and (d), respectively). Note that we have since developed the capability to use the 'singular' integration scheme for arbitrary discontinuous discretisations. The new scheme allows a further reduction in modelling time while maintaining computational efficiency; it is a part of EZBEA 2.3.

9.2.13 Combining BEA and FEA

As stated in the foregoing, BEA has cost and accuracy advantages but lacks the capability required for some models. The combination of BEA and FEA seems obvious and has been done in two ways at Caterpillar.

9.2.13.1 Structural Zooming

Structural zooming uses the results of one model (generally FE) as the boundary conditions for a subsequent model (generally BE). It can be used effectively to evaluate results near stress raisers such as drilled holes and machined radii. It can also be used for a quick reanalysis of part of a structure as an aid in understanding laboratory and field failures. Implicit to the method is the assumption that the original model is adequate to model the results in 'nominal' areas.

The authors (Graf and Gebre-Giorgis, 1986) have described a program which automatically interpolates a FE model's results and computes the boundary conditions for a refined model. This program was used on the cylinder head example in the following section.

9.2.13.2 Coupled BEA and FEA

A second method of combining FEA and BEA is to couple the FE and BE models and solve them together (see, for example, Rudolphi, 1985). EZBEA 2.2 has the capability of generating a stiffness matrix that can be combined with an FE model using MSC/NASTRAN®. The authors (Graf and Gebre-Giorgis, 1986) have described the method used to generate the stiffness matrix. While the required user input for EZBEA is modest, the computer time for large problems is not (many right-hand-side vectors must be processed). Additionally, the user must run both MSC/NASTRAN and EZBEA (the latter twice) to solve his model. Finally, we have not resolved the decreased accuracy of the three-dimensional BEA-generated stiffness matrix compared to comparable BEA or FEA models. Consequently, we have not utilised the coupled BEA/FEA approach for large problems.

9.3 EZBEA EXAMPLES

In this section, we will present results for various example problems solved using EZBEA. Most of these problems are actual machine components analysed by design engineers at Caterpillar. We had little input into how these models were built—some meshes are much more refined than we believe to be necessary. These problems are intended to show the versatility of EZBEA in solving various types of practical problems. The model sizes and CPU times required to solve.the models on various computers are summarised in Table 4.

TABLE 4

SUMMARY OF EZBEA EXAMPLES—MODEL SIZES AND CPU TIMES

Model name	Analysis type[a]	Number of			Computer used	CPU time(s)
		Subreg.	Elem.	Nodes		
Hollow cyl. I	3D HC	1	6	29	CDC 855	8
Hollow cyl. II	3D HC	1	24	93	CDC 855	28
Hollow cyl. III	3D HC	1	200	651	CDC 855	733
Hollow cyl. I	3D TE	1	6	29	CDC 855	31
Hollow cyl. II	3D TE	1	24	93	CDC 855	153
Hollow cyl III	3D TE	1	200	651	CDC 855	8 607
Fuel inj. tip	3D EL	2	259	828	IBM 3090	1 051
U-joint	3D EL	1	172	517	VAX 8530	7 467
Spiral pinion	3D EL	1	545	1 621	VAX 8800	68 614
Spiral pinion	3D EL	1	545	1 621	FPS 364	31 719
Spiral pinion	3D EL	1	545	1 621	FPS 364[b]	17 119
Spiral pinion	3D EL	1	545	1 621	FPS 264	17 461
Spiral pinion	3D EL	1	545	1 621	IBM 3090	15 471
Spiral pinion	3D EL	1	545	1 621	IBM 3090VF[c]	2 199
Crankshaft	3D EL	3	541	1 459	IBM 3090	4 172
Cylinder head	3D EL	1	196	586	IBM 3090	851
Pivot shaft	AX EL	1	174	349	VAX 8530	383

[a] HC—heat conduction, TE—thermoelastostatic, EL—elastostatic.
[b] With FMSLIB® (Fast Matrix Solution Library).
[c] With ESSL® (Engineering and Scientific Subroutine Library).

9.3.1 Hollow Cylinder

The following example demonstrates the accuracy of EZBEA for a wide range of model sizes. The model is of a hollow cylinder with the ends restrained and perfectly insulated. The inner surface is maintained at a uniform temperature of $0 \cdot 0°$; the outer surface has a film coefficient of $0 \cdot 5$ and a bulk temperature of $5 \cdot 35°$. Figure 7 shows the three models used in

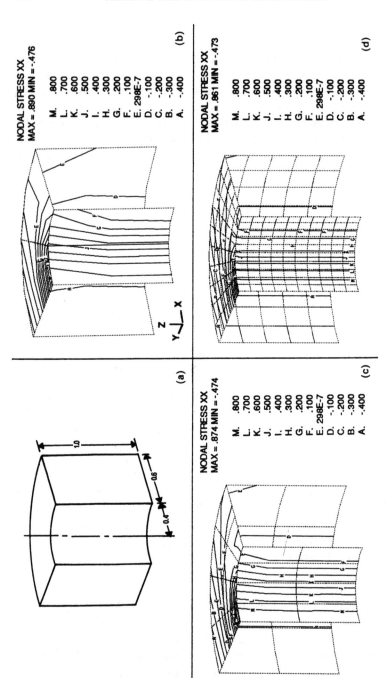

FIG. 7   Three-dimensional EZBEA hollow cylinder models.

this example. The first model has two elements each on the inside and outside surfaces and two elements on the top. Though these elements are barely sufficient to define the geometry, the computed temperature was in error by only 0·1%. The subsequent thermoelastic analysis predicted radial displacements on the outer surface that were in error by only 0·2%. Figure 7 shows the three models and the thermal stresses (σ_{xx}) each predicted.

9.3.2 Fuel Injector Tip

In this case, the fuel injector tip shown in Fig. 8 is to be analysed. Because of the complicated geometry, problems such as this are well suited to BEA.

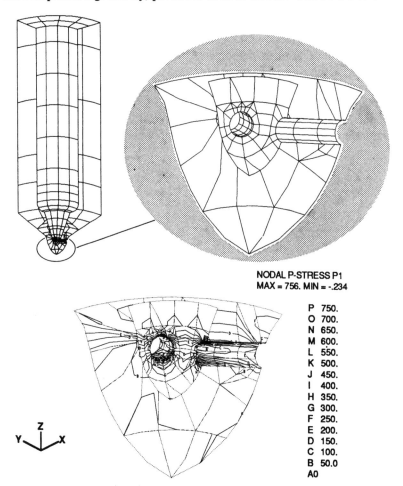

NODAL P-STRESS P1
MAX = 756. MIN = -.234

P	750.
O	700.
N	650.
M	600.
L	550.
K	500.
J	450.
I	400.
H	350.
G	300.
F	250.
E	200.
D	150.
C	100.
B	50.0
A	0

FIG. 8   Fuel injector tip model and stresses.

The overall length of the injector tip is 19 mm while the diameter is 10 mm. There are six injection ports with diameter of 0·2355 mm. Loading is due to fuel pressure inside the injector. Because of symmetry, only one-quarter of the injector tip was modelled. The top surface was constrained in the axial direction to prevent rigid body motion.

The BEA model used to solve the problem is shown in Fig. 8. Subregion 1, which is the lower tip area, contains 175 elements. Subregion 2, which makes up the remaining portion of the part, contains 92 elements. Because the dimensions of the problem are small, the elements at the tip can be seen only in the magnified view in Fig. 8. Also shown is a close-up view of the maximum principal stresses around the injector ports.

9.3.3 Universal Joint

This example is used to demonstrate the program's ability to analyse problems with symmetric geometry and symmetric or antisymmetric boundary conditions. The problem illustrated in Fig. 9 involves a universal joint (U-joint) used on a track-type loader. The U-joint model was loaded at the pin bore as shown in Fig. 9. The objective was to determine the maximum deflection and stress in the component.

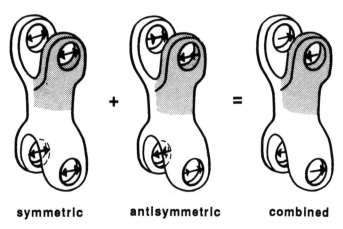

symmetric   antisymmetric   combined

FIG. 9   Universal joint torsional loading (modelled region shaded).

In order to do this, a model of one-fourth of the U-joint was created as shown in Fig. 10(a). Two separate runs—one symmetric and the other antisymmetric—were required to solve the problem. (A one-eighth model could have been used but more runs would have been required.) Figure

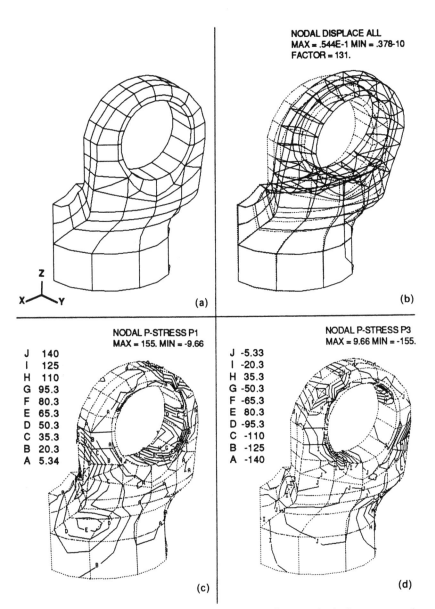

FIG 10   Universal joint model, displacements, maximum principal stresses, and
minimum principal stresses.

9 summarises the loading used in the two runs. In the symmetric loading, the opposite ends of the inside of the pin bores were loaded with one-half the desired normal traction. The results of this run were saved in EZBEA's restart FILE1. In the antisymmetric run, the other half of the loading was applied to the same pin bore surfaces. The tractions used to cancel the unwanted tractions applied in the symmetric load case were negative (tensile) pressures. During the post-processing of the second run, the results of the symmetric and antisymmetric load cases were combined to form the desired torsional loading.

The predicted deformed shape is shown in Fig. 10(b). The contour plots of the maximum and minimum principal stresses are shown in Fig. 10(c) and (d).

9.3.4 Spiral Bevel Pinion
This example (Fig. 11) is a bevel pinion stress analysis that was done to verify a simpler analysis procedure. The pinion is 50 mm long and has an outside diameter of 100 mm. The tooth was loaded with parabolically varying pressure along a diagonal strip on the face of the tooth. This represents the contact with the mating gear.

Figure 11(b) shows the boundary element model of the pinion. To approximate the entire pinion, simple representations of the teeth on either side of the loaded tooth were included. The ends of these teeth and the inside surface of the pinion were constrained in the normal direction. The objective of the analysis was to determine the maximum stress at the root of the loaded tooth. Maximum and minimum principal stress contours are shown in Fig.11(c) and (d).

9.3.5 Crankshaft
One throw of a diesel engine crankshaft (see Fig. 12) was modelled using EZBEA, including one rod journal and half of the main journals on either side of the rod journal. This three-subregion model included details in the journal fillets and the rod journal oil holes.

The crankshaft is subjected to torsional loading. The BEA model shown in Fig. 12 was loaded using tangential tractions on one end of the main journal while the opposite end was constrained. The applied traction was zero at the centre of the journal and a maximum at the outside surface. Figure 12(b), (c) and (d) shows the stress results predicted by EZBEA. The stresses around the oil hole, a critical location, are shown in Fig. 12(c) and (d). As expected, the maximum and minimum stresses are at 90° intervals around the hole. Note that the maximum stress area is located inside the oil

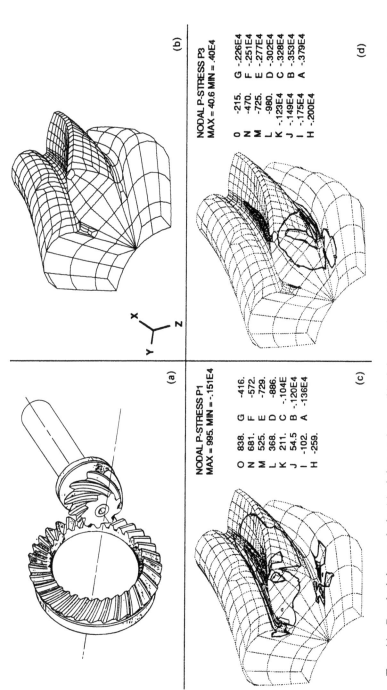

(b)

NODAL P-STRESS P3
MAX = 40.6 MIN = -.40E4

O -215.   G -226E4
N -470.   F -251E4
M -725.   E -277E4
L -980.   D -302E4
K -.123E4   C -328E4
J -.149E4   B -353E4
I -.175E4   A -379E4
H -.200E4

(d)

(a)

NODAL P-STRESS P1
MAX = 995. MIN = -.151E4

O 838.   G -416.
N 681.   F -572.
M 525.   E -729.
L 368.   D -886.
K 211.   C -.104E4
J 54.5   B -.120E4
I -102.   A -.136E4
H -259.

(c)

FIG. 11 Bevel pinion analysis: (a) pinion and gear; (b) pinion tooth model; (c) and (d) maximum and minimum principal stresses.

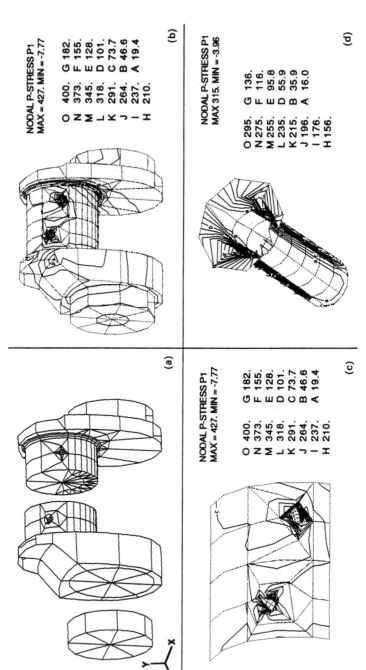

NODAL P-STRESS P1
MAX = 427. MIN = -7.77

O 400. G 182.
N 373. F 155.
M 345. E 128.
L 318. D 101.
K 291. C 73.7
J 264. B 46.6
I 237. A 19.4
H 210.

(b)

NODAL P-STRESS P1
MAX 315. MIN = -3.96

O 295. G 136.
N 275. F 116.
M 255. E 95.8
L 235. D 55.9
K 215. B 35.9
J 196. A 16.0
I 176.
H 156.

(d)

(a)

NODAL P-STRESS P1
MAX = 427. MIN = -7.77

O 400. G 182.
N 373. F 155.
M 345. E 128.
L 318. D 101.
K 291. C 73.7
J 264. B 46.6
I 237. A 19.4
H 210.

(c)

Fig. 12   Crankshaft model and maximum principal stresses (overall view and detail near oil hole).

hole *below* the journal surface. These results were verified with laboratory measurements and were reported by Huber and Sanders (1987).

9.3.6 Cylinder Head

Engine cylinder heads are analysed by constructing large FEA models of the head, engine block, and cylinder liner. FEA is used primarily because these components have high surface-to-volume ratios; BEA models of these structures would require extremely high computing resources. In order to utilise some of the advantages of BEA, a coarse FEA model is constructed which will accurately represent the overall stiffness of the model. The displacements of this model are used as boundary conditions for a refined grid submodel in the area of interest. In this example, structural zooming was used to guide the redesign of the intake port of an engine cylinder head. The coarse FEA model of the cylinder head shown in Fig. 13 was constructed to represent the stiffness of the structure. The displacements calculated in the port area were used as boundary conditions to the refined BEA model which is shown in Fig. 13(c). A program called ZOOM was used to interpolate the FE model's displacements at each node in the BE model that required displacement boundary conditions.

Because of the modelling simplicity of BEA, various modifications were tried in a relatively short time to reduce the stresses of the intake area. The results of the final design are shown in Fig. 13. The maximum principal stresses, shown in Fig. 13(d), are 27% lower than the stress in the initial design. It is important to note that the accuracy of the results of the 'zoomed' model depends on the accuracy of the coarse FEA model which predicted the displacement boundary conditions.

9.3.7 Pivot Shaft

This example problem is intended to demonstrate EZBEA's capability to analyse axisymmetric models with non-axisymmetric loading. The part is a track-type loader pivot shaft. The model is loaded on two bearing surfaces in constant, cosine Θ, and cosine 2Θ modes are shown in Fig. 14. This combination of load will represent a bearing-type loading. The shaft is constrained at the flange area where it is bolted to the main frame as well as the inside surface of the tubular section where it fits into the main frame. As can be seen from the model, no elements are required along the axis of symmetry.

The displacements predicted using the pivot shaft model are shown in Fig. 15, as are the principal stresses along the three radii in the model as labelled.

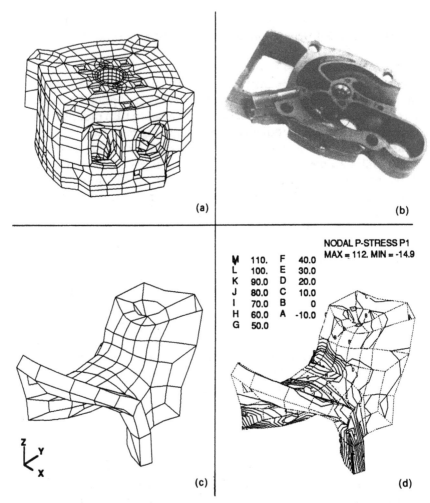

FIG. 13   Coarse FE cylinder head model, cross-section through the failed cylinder head, EZBEA cylinder head submodel, and maximum principal stresses.

9.4 FUTURE DEVELOPMENTS

There is ongoing development and maintenance of EZBEA at Caterpillar. This development still emphasises program reliability, portability, efficiency, and compatibility. Future developments are likely to include user-specified linear constraint equations and 'gap' elements (unilateral

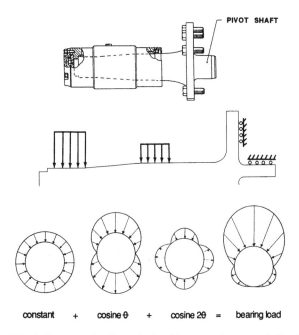

FIG. 14   Axisymmetric pivot shaft with non-axisymmetric loading.

constraint equations). These will facilitate modelling of assemblies of complex components which have either determinate or indeterminate contact.

We also plan to develop further the ability to couple FE and BE models. At a minimum, we plan to resolve the decreased accuracy of BEA-generated stiffness matrices for three-dimensional models. We also hope to reduce the required resources—both human and computational. Human resources can be saved by streamlining the procedure required to solve a coupled model. This may be done by better integration of separate FE and BE programs or by providing both FE and BE capabilities in a single program. Computational resources can be saved by improved computational efficiency.

A longer-range development area is to integrate EZBEA with a solid modeller, automatically generating the BE model, solving it, and refining it based on the initial results. We have considerable experience with this general approach for two-dimensional models; we have lacked the geometric definition required for three-dimensional models.

To support the 'gap' element capability, an iterative solution will be

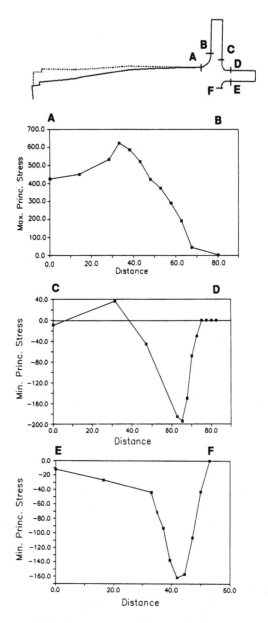

FIG. 15   Pivot shaft displacements, maximum principal stresses, and minimum principal stresses.

required. It seems that an iterative solver could also be used for automatic mesh refinement—the solution for the previous level of mesh refinement provides a good approximation of the solution for the current mesh. This could be sufficient to make automatic mesh refinement computationally feasible today.

9.5 CONCLUSION

Boundary element analysis is applied routinely at Caterpillar, especially to problems with complex geometry. The method is used almost exclusively for some components. In the long term, we expect we will see boundary elements and finite elements used together in a single model, solved by a single general-purpose analysis program.

REFERENCES

ANON. (1986). *EZBEA Verification Examples Manual*, Research-927, Caterpillar, Inc., Peoria, Illinois.

COX, J.V. & SHUGAR, T.A. (1985). A recursive integration technique for boundary element methods in elastostatics, *Proc. ASME Conf. on Advanced Topics in Boundary Element Analysis*, Eds T.A. Cruse, A.B. Pifko and H. Armen, AMD Vol. 72, ASME, New York.

GERSTLE, W.H. & INGRAFFEA, A.R. (1985). Error control in three-dimensional crack modeling using the boundary element method, *Proc. ASME Conf. on Advanced Topics in Boundary Element Analysis*, Eds T.A. Cruse, A.B. Pifko and H. Armen, AMD Vol. 72, ASME, New York.

GRAF, G.L. & GEBRE-GIORGIS, Y. (1986). Combining finite element and boundary element analyses, *SAE paper 860747*, SAE, Warrendale, PA.

HOLZE, G.H. (1980). Boundary integral equation method simplifies elastic stress analysis, *SAE paper 800431*, SAE, Warrendale, PA.

HUBER, R.W. & SANDERS, R.J. (1987). The application of boundary element analysis to engine component design, *SAE paper 870578*, SAE, Warrendale, PA.

LACHAT, J.C. (1975). A further development of the boundary integral technique for elastostatics, Ph.D. Thesis, University of Southampton, UK.

LACHAT, J.C. & WATSON, J.O. (1976). Effective numerical treatment of boundary integral equations: a formulation for three-dimensional elastostatics, *Int. J. Numer. Methods in Eng.*, **10**, 991–1005.

MUKHERJEE, S. & MORJARIA, M. (1984). On the efficiency and accuracy of the boundary element method and the finite element method, *Int. J. Numer. Methods Eng.*, **20**, 515–22.

NIGAM, R.K. (1980). The boundary integral equation method for elastostatics problems involving axisymmetric geometry and arbitrary boundary conditions, M.S. Thesis, University of Kentucky.

RIZZO, F.J. & SHIPPY, D.J. (1977). An advanced boundary integral equation method for three-dimensional thermoelasticity, *Int. J. Numer. Methods Eng.*, **11**, 1753–68.

ROARK, R.J. & YOUNG, W.C. (1975). *Formulas for Stress and Strain*, 5th edn, McGraw-Hill, New York.

RUDOLPHI, T.J. (1983). An implementation of the boundary element method for zoned media with stress discontinuities, *Int. J. Numer. Methods Eng.*, **19**, 1–15.

RUDOLPHI, T.J. (1985). Nonhomogeneous potential and elasticity problems by combined boundary and finite elements, *Proc. ASME Conf. on Advanced Topics in Boundary Element Analysis*, Eds T.A. Cruse, A.B. Pifko and H. Armen, AMD Vol. 72, ASME, New York.

SHIGLEY, J.E. (1972). *Mechanical Engineering Design*, 2nd edn, McGraw-Hill, New York.

WATSON, J.O. (1979). Advanced implementation of the boundary element method for two- and three-dimensional elastostatics. Chapter 3 in: *Developments in Boundary Element Methods–1*, Eds P.K. Banerjee and R. Butterfield, Applied Science Publishers, London, 31–63.

INDEX

Acoustic eigenfrequency analysis
  application of GPBEST, 218–23
  BEM formulation for, 214–18
Automotive industry, 205–229
  analysis of structural components, 223–7
  BEM experience, 209–12
  road boom conceptual design, 212–23
Axisymmetric steel pressure vessel, 62–4

BEASY code, 210–11, 224
Benchmark notch specimen, 19
  finite-element meshes for, 24
  under monotonic loading, 25
BEST3D code, 16, 17, 19, 26–35, 84, 210–11
BETSY code, 233–4
BETTI program code, 257–9
Bevel gear toothing, BE model, 253–4
Biot's theory, 121, 122
Boundary element method (BEM), mathematical theory, 207–9
Boundary integral equation method (BIEM), 158
Buried caisson foundation, 108–12

Caisson foundation, 108–12
Camshaft, BE models, 241–50
Cauchy's principal value, 162
CDC CYBER176, 224
Centrifugal loading, 6

Circular crack, BEM map, 195–6
C-notch low cycle fatigue specimen, 17, 18
Complementary function, 52, 59
Consolidation
  Koyna Dam, at, 143–6
  rectangular loads, under, 146–8
  strip load, under, 139–43
Continuum mechanics problems, 205
Cooling of steel sphere, 148–50
Coupled Quasistatic Thermoelasticity (CQT), 127–30
Crack analysis, 182–3
Crack opening displacement, 183, 196
Crack tip behaviour, 188
Crack tip elements, 192
Cracked bodies, BEM modelling, 191–2
Crankshaft, EZBEA, 286
Crankshaft throw, BE model, 250–2
Cray-1 computer, 69
CRAY-XMP computer, 104, 221, 225
CRX3D computer code, 198–9, 201
Cylinder head, EZBEA, 289
Cylindrical rod, residual stresses, in 59–62

Darcy's law, 124
DBETSY program system, 231–61, 234–41
  comment on, 256
DBETSY 1.0 program system, 234–6
  applications, 241–54
  comment on, 256

DBETSY-3D program system,
    236–9
  applications, 241–54
  blocked residual matrix technique,
    239–41
  calculated volume integrals, 238
  comment on, 256
Discretised boundary integrals, 194–5
Double edge notch specimen, 19
Duhamel–Neumann constitutive laws,
    126
Dynamic plasticity, 87–93
  constitutive model, 89–90
  numerical implementation, 90–1
  stresses at boundary points, 89
  stresses at interior points, 88–9
  time-marching scheme, 91–3
Dynamic reciprocal theorem, 82

Elastic analysis with body forces, 4–5
Elastic dynamic analysis of solids,
    77–117
Elastic half-space, 104–8, 110
  surface response of, 112–15
Elastic properties, definition of, 3
Elastodynamic problems, 158–9
  reduced, 159–60
Elastodynamic wave propagation
    problems, 157–86
  BIE formulations problems, 161–71
  BIE–FEM method, 171–8
  boundary stresses, 175
  crack problems, 182–3
  discretisation of BIE, 174
  domain-typed integral equations,
    162–8, 178–80
  examples, 175–83
  Fourier transformed domain,
    161–2
  numerical techniques, 171–5
  regularisation of hypersingular
    kernels, 168–71
  surface potentials, 164
  three-dimensional problem, 165–7
  time domain, 162
  two-dimensional problem, 167–8
  uniqueness of solutions, 175

Elastodynamic wave propagation
    problems—contd.
  volume potentials, 164–5
Elastoplastic analysis, axisymmetric
    steel pressure vessel, 62–4
Elastoplastic flow equations, 41–2
Elastoplastic integral formulation,
    42–59
Elliptical surface crack, 196–9
Elliptical surface flaw, 199–201
EZBEA, 212, 266–9
  development issues, 266–8
  efficiency, 268
  examples, 281–9
  future developments, 290–3
  heat conduction problems, 268–9,
    271–3, 273
  model sizes and CPU times, 281
  multiple subregions, 271–6
  non-axisymmetric loading of
    axisymmetric models, 276–8
  portability, 267–8
  reliability, 266–7
  thermoelasticity capability, 270
  thermoelasticity example, 273–6
  thermoelasticity, implementation,
    272–3
EZBEA code, 210–11, 280–90
EZBEA 2.2
  advanced numerical integration,
    278–80
  coupled BEA and FEA, 280
  summary of capabilities, 267
EZBEA 2.3, 280

FEMVIEW, 268
Finite difference method, 205
Finite element method (FEM), 69,
    158, 171–4, 206, 209, 213,
    214, 264–5
Flexible square plate formulation,
    104–6
Fourier inversion, 174
Fourier series, 82
Fourier transform, 159–60
Fracture mechanics, 187–204
  BEM analysis, 190–1

Fracture mechanics—*contd.*
numerical solutions, 195–201
Free-vibration analysis, 8–12
twisted plate, of, 29–32
Fuel injector tip, EZBEA, 283–4

Galerkin vector, 52–3
Gas turbine engine structures, 1–37
BEM formulations, 4–12
component analysis, 26–9
free–vibration of twisted plate,
29–32
geometrical complexity, 2
loading definition, 3
material property definition, 3–4
numerical implementation, 12–17
problem sizes, 2
General axisymmetric formulation,
80–2
Generic modelling regions (GMRs),
80, 139, 150
Global shape function, 54–6
GPBEST code, 84, 106, 138, 148, 151,
210–11, 214, 218, 225, 227
Gravitational loading, 5

Heat conduction, 268–9, 271–3, 273
Hewlett-Packard 9000 computer, 71
Hollow cylinder, EZBEA, 281–3
Hooke's Law, 53

Inclined circular crack, 199
Inelastic analysis, 6–9
dynamic
solids, of, 77–117
numerical implementation,
79–80
solids, of, 39–75
numerical implementation, 45–6,
56–8
uniform and weighted mesh for, 22
Inelastic response, 4
cyclic loading, under, 26
monotonic loading, under, 23
Infinite plate under triaxial stress, 189

Initial stress expansion technique,
46–7
Initial stress rate, 56, 57, 59

Koyna Dam, consolidation at, 143–6

Lamé constant, 129
Lamé modulus, 129
Laplace's equation, 268
Layered soil medium, 93–5
Linear elastic facture mechanics
(LEFM) analysis, 188–9
LINPAC, 80

Machine components, thermoelastic
analysis, 263–94
MHOST computer code, 24
MSC/NASTRAN, 209, 212, 225, 227,
280

Navier–Cauchy equations, 82
Nodal density rate, 59
Notch plate, three-dimensional
analysis, 64–6

Particular integrals, 52–4, 56, 57, 59
PATRAN, 212
Perforated plate
three-dimensional analysis, 66–70
two-dimensional analysis, 70–1
Periodic dynamic analysis, 78–82
Pivot shaft, EZBEA, 289
Plastic deformation
notch plate, 64
perforated plate, 66–70
Poroelastic analysis, 119–56
applications, 139–53
boundary integral formulation,
134–8
fundamental solutions, 130–4
numerical implementation, 138–9
three-dimensional, 130–2
two-dimensional, 132–4

Poroelastic constants, 125
Poroelastic–thermoelastic analogy, 128–30
Poroelasticity, governing equations, 122–6
Primary loading modes of cracked body, 188
Product development, 263–4

Quarter point elements (QP), 196

Rectangular loads, consolidation under, 146–8
Residual stresses in cylindrical rod, 59–62
Ricker's wavelet, 180
Riemann convolution, 88
   integrals, 83
Rigid circular disc, vibration analysis, 106–7
Road boom design quality, 212–23

Self-adaptive integration scheme, 14
Semi-elliptic surface crack, 198
Software requirements, 265
Soil-structure interaction problems, 115
Solids, inelastic analysis of. See Inelastic analysis
Spiral beval pinion, EZBEA, 286
Steel sphere, cooling of, 148–50
Stokes' solution, 82
Stress intensity factor, 182, 189, 190, 193–4, 196, 198, 199, 201–3
Strip load, consolidation under, 139–43
Substructured analysis, 157–86
Subsurface cuboidal region, 112–15
SUPERTAB, 268
Surface foundation displacements, 175–6
Surface response of elastic half-space, 112–15

T-joint section with elliptical surface flaw, 199–201

T-shaped bridge pier response, 176–8
Thermal diffusivity, 128
Thermoelastic analysis, 119–56
   applications, 139–53
   BEA vs FEA, 264–5
   boundary integral formulation, 134–8
   combining BEA and FEA, 280
   fundamental solutions, 130–4
   machine components, 263–94
   numerical implementation, 138–9
   three-dimensional, 130–2
   two-dimensional, 132–4
Thermoelasticity, governing equations, 126–8
Time-domain boundary element formulation, 121
Time-domain coupled problem, 121
Time-integrated total normal mass flow, 122
Time-marching scheme, 91–3
Traction singular (TS) modification, 196
Transient boundary integral formulation, 82–7
   numerical implementation, 84–6
Turbine blade plane strain response, 150–3

Underground explosion, 93–5
Universal joint, EZBEA, 284–6

Variable stiffness elastoplasticity
   solution process for, 50–1
   system equations for, 49–50
Variable stiffness plasticity, formulation, 44–5
Vibration analysis, rigid circular disc, 106–7
Vibration isolation problem, 96–104
Viscoelastodynamic problems, 160–1
Volume integral formulations, 42–4
Von Mises yield criterion, 59, 63, 64, 67
Von Mises yield surfaces, 25

Milton Keynes UK
Ingram Content Group UK Ltd.
UKHW020022071024
449327UK00032B/2883